MATH PHONICS™

FRACTIONS

Quick Tips and Alternative Techniques for Math Mastery

BY MARILYN B. HEIN

ILLUSTRATED BY RON WHEELER

Teaching & Learning Company

1204 Buchanan St., P.O. Box 10
Carthage, IL 62321-0010

THIS BOOK BELONGS TO

_____

ACKNOWLEDGEMENTS

Sincere thanks to Kay Suchan for being my consultant on this book; to Nick Hein for ideas for games and challenge problems and Jennifer Hein for technical support.

DEDICATION

I would like to dedicate this book to my husband, Joe; our children, Gretchen, Troy, Adam, Sarah, Robert, Nick and Jenny; and my parents, Vincent and Cleora Vestring.

Cover by Ron Wheeler

Copyright © 1998, Teaching & Learning Company

ISBN No. 1-57310-112-5

Printing No. 98765432

Teaching & Learning Company
1204 Buchanan St., P.O. Box 10
Carthage, IL 62321-0010

Math Phonics™ is a trademark registered to Marilyn B. Hein.

TABLE OF CONTENTS

NOTE TO PARENTS

Although this book has been specifically designed to be used by classroom teachers for teaching fractions, the materials are extremely helpful when used by parents and children at home.

If you have purchased this book to use at home with your child, I recommend that all of the fill-in-the-blank pages be inserted into vinyl page protectors and worked with a dry-erase overhead transparency marker. The page protectors can be washed and the page can be reused. Put all the page protectors and worksheets into a vinyl, two-pocket binder. The pen, flash cards and other materials can be kept in the binder pockets, thus creating a handy, portable math kit.

I think you will find that these methods work extremely well both at home and in the classroom. It's a book you can count on!

Dear Teacher or Parent,

Fraction concepts are quite easy to understand, but some people (young and old) are totally confused by them! One reason is that students are weak in the basic math facts. They get lost while working a fraction problem with several steps. Another reason is that students don't understand basic fraction concepts. Though they may be able to do today's lesson, they forget very quickly because the information has not really been learned.

Math Phonics™ is a comprehensive approach to teaching basic math concepts. As teaching reading through phonics helps children break down unrecognizable words into recognizable phonetic components, so *Math Phonics™* teaches children (and adults!) that the best way to handle a confusing math problem is to break it down into recognizable basic facts.

This book introduces fractions with basic methods of computation (no carrying, borrowing or cancelling). There are 10 easy-to-follow lesson plans, worksheets, assessment pages and more—all based upon the *Math Phonics™* principles.

You'll find simple classroom demonstrations and manipulatives designed to help students find answers and better understand concepts. Also, wall charts are included which provide visual reinforcement of important ideas. Simple, at-home games are included so students can practice without boredom.

Math Phonics™ ideas have been tried and approved in homes and classrooms and they **work**!

So take a look! I have included everything in this book that I have found helpful in introducing fractions during my years as a teacher, substitute teacher, tutor and parent. I love learning new things. I love teaching new things, and I truly believe that students can learn to love learning new things provided they have some fun in school once in a while—even with fractions!

Sincerely,

Marilyn

Marilyn B. Hein

WHAT IS MATH PHONICS™?

Math Phonics™ is a specially designed program for teaching fractions or for remedial work.

WHY IS IT CALLED MATH PHONICS™?

In reading, phonics is used to group similar words, and it teaches the students simple rules for pronouncing each word.

In *Math Phonics*™, math concepts are learned by means of simple patterns, rules and wall charts with games for practice.

In reading, phonics develops mastery by repetitive use of words already learned.

Math Phonics™ uses drill and review to reinforce students' understanding. Manipulatives and practice games help reduce the "drill and kill" aspect.

HOW WAS MATH PHONICS™ DEVELOPED?

Why did "Johnny" have so much trouble learning to read during the years that phonics was dropped from the curriculum of many schools in this country? For the most part, he had to simply memorize every single word in order to learn to read, an overwhelming task for a young child. If he had an excellent memory or a knack for noticing patterns in words, he had an easier time of it. If he lacked those skills, learning to read was a nightmare, often ending in failure—failure to learn to read and failure in school.

Phonics seems to help many children learn to read more easily. Why? When a young child learns one phonics rule, that one rule unlocks the pronunciation of dozens or even hundreds of words. It also provides the key to parts of many larger words. The trend in U.S. schools today seems to be to include phonics in the curriculum because of the value of that particular system of learning.

As a substitute teacher, I have noticed that math teachers' manuals sometimes have some valuable phonics-like memory tools for teachers to share with students to help them memorize fraction concepts. However, when I searched for materials which would give students a chance to practice and review these tools, I found nothing like that available. I decided to create my own materials based upon what I had learned during the past 40 years as a student, teacher and parent.

The name *Math Phonics*™ occurred to me because the rules, patterns and memory techniques that I have assembled are similar to language arts phonics in several ways. Most of these rules are short and easy to learn. Children are taught to look for patterns and use them as "crutches" for coming up with the answer quickly. Some concepts have similarities so that learning one makes it easier to learn another. Last of all, *Math Phonics*™ relies on lots of drill and review, just as language arts phonics does.

Children *must* master basic fraction concepts and the sooner the better. When I taught seventh and eighth grade math over 20 years ago, I was amazed at the number of students who had not mastered fractions. At that time, I had no idea how to help them. My college math classes did not give me any preparation for that situation. I had not yet delved into my personal memory bank to try to remember how I had mastered those basics.

When my six children had problems in that area, I was strongly motivated to give some serious thought to the topic. I knew my children had to master the basics, and I needed to come up with additional ways to help them. For kids to progress past the lower grades without a thorough knowledge of those concepts would be like trying to learn to read without knowing the alphabet.

I have always marveled at the large number of people who tell me that they "hated math" when they were kids. I firmly believe that a widespread use of *Math Phonics*™ could be a tremendous help in solving the problem of "math phobia."

WHAT ARE THE PRINCIPLES OF MATH PHONICS™?

There are three underlying principles of *Math Phonics*™.
They are: 1. Understanding
            2. Learning
            3. Mastery
Here is a brief explanation of the meaning of these principles.

1. **UNDERSTANDING:** All true mathematical concepts are abstract which means they can't be touched. They exist in the mind. For most of us, understanding such concepts is much easier if they can be related to something in the real world—something that can be touched.

   Thus, I encourage teachers to let students find answers for themselves using fraction strips and circles. I think this helps the students to remember answers once they have discovered them on their own.

2. **LEARNING:** Here is where the rules and patterns mentioned earlier play an important part. A child can be taught a simple rule and on the basis of that, begin to practice with fractions. But the learning necessary for the basic fraction concepts must be firmly in place so that the information will be remembered next week, next month and several years from now. That brings us to the next principle.

3. **MASTERY:** We have all had the experience of memorizing some information for a test or quiz tomorrow and then promptly forgetting most of it. This type of memorization will not work for fractions. In order for children to master fractions, *Math Phonics*™ provides visual illustrations, wall charts, manipulatives, worksheets and games. Some children may only need one or two of these materials, but there are plenty from which to choose for those who need more.

TLC10112 Copyright © Teaching & Learning Company, Carthage, IL 62321-0010

POCKET FOLDERS

You will want to purchase or create a pocket folder for each student to keep all the *Math Phonics*™ materials together.

Inexpensive pocket folders are available at many school or office supply stores, discount stores or other outlets.

An easy-to-make pocket folder can be made from a large paper grocery or shopping bag.

1. Cut away the bottom of the bag and discard.

2. Cut open along one long side and lay flat.

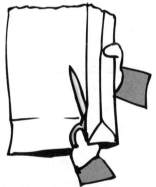

3. Pick one of the folds and measure out 10" (25 cm) from the fold on either side. Trim bag.

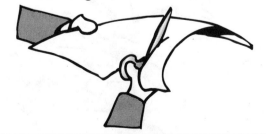

4. Now measure 12" (30 cm) down from the top and fold up the remaining portion of the bag.

5. Staple pockets at outside edges and fold in half.

6. Decorate front and back.

Suggest to parents that children should keep all of their *Math Phonics*™ materials (worksheets, fraction strips, wall charts, etc.) in this folder. Parents may also wish to supply clear plastic page protectors and dry-erase markers. Worksheets can be inserted into the page protectors, completed with the dry-erase marker and reused.

SUMMARY OF THE 10 BASIC STEPS

1. Understanding Fractions

This step is important even if students have worked with fractions in the past. One friend told me he didn't really understand the base 10 system until he was in college. Lots of people go through school hating math because they don't understand basic concepts.

2. Simple Addition and Subtraction of Fractions

When I say simple, I mean simple! This step has no changing denominators, no carrying, no borrowing, no improper fractions, no mixed numbers and no reducing. Reviews of addition and subtraction of whole numbers are included.

3. Equivalent Fractions

Equivalent fraction strips are provided for classroom demonstrations of this concept. Multiplication review is included.

4. Reducing Fractions

These are also equivalent fractions written in lowest terms. Division review is included.

5. Finding Common Denominators for Addition and Subtraction

In this step we will cover the two simplest ways for finding common denominators—using the larger denominator as the common denominator, ($1/2$ + $1/4$—4 is the common denominator) and multiplying the two denominators to find the common denominator ($1/3$ + $1/5$—15 is the common denominator). Finding common denominators for larger numbers will be taught in step 10.

6. Improper Fractions

This is another way of reducing. Students will learn to change a fraction like $6/3$ to its whole number value of 2, or $10/3$ to its mixed number value of $3 1/3$.

7. Mixed Numbers

This is the flip side of step 6. Here students will learn to change a mixed fraction like $2 1/2$ to its improper fraction $5/2$. Fraction circles can be used for classroom demonstrations.

8. Multiplication of Fractions

This step will again use equivalent fraction strips to demonstrate to students that multiplication of fractions really gives the right answer.

9. Division of Fractions

This lesson will relate directly to the multiplication of fractions.

10. Finding Least Common Denominators

This step involves prime factorizations and common multiples.

Rules, Games and Assessment

Includes a summary of rules and games. Assessment pages are included.

INTRODUCTION: Although your students have probably worked with fractions in the past, spend a few minutes explaining the idea of a fraction. Here's why.

1. Some students may have completely missed fractions in the past due to changing schools or being sick.

2. Some students may have forgotten what was said earlier about fractions.

3. Some students may have done problems with fractions automatically and not understood what they were doing. When that is the case, students quickly forget which method to use with addition and which one to use with multiplication. With each year of life, students are more capable of understanding concepts, and it is important to repeat explanations from time to time.

THE IDEA OF A FRACTION: A fraction is a way of writing a number. A fraction means to divide. It shows part of an object or part of a group of objects.

Students already know several ways of writing numbers. Ask your class how many different ways they can think of to write the number one. We have the word *one*, the numeral *1*, the Roman numeral *I*, and we can draw one object to represent the number one. We could do the same with the number 10, the number 5000 and so on.

Sometimes we take an object (like an apple pie) and cut it into equal pieces, and we need a way to talk about one of the pieces. It we cut it into 4 pieces, we write the fraction $1/4$. Show the class the fraction $1/4$ and tell them the 4 below the line means we cut it into 4 equal pieces, and the 1 above the line means we are talking about one of the pieces. Draw a circle on the board and divide it into fourths. Explain the fraction $2/4$, the fraction $3/4$ and the fraction $4/4$. Then draw another circle and demonstrate the fraction $1/3$, the fraction $2/3$ and the fraction $3/3$. Be sure they know the pieces must be the same size. The number above the line is called the numerator, and the number below the line is called the denominator.

RULE: A fraction is a number which shows that something has been divided. It has a numerator above the line and a denominator below the line.

PRACTICAL USES: Have students think of two or three ways that people might use fractions in their jobs or their homes. (Recipes, measuring wood for a building project, cutting fabric, sharing a candy bar, getting the right wrench to fix a bike, etc.)

WALL CHART: Post the Numerator-Denominator Wall Chart A (page 13) in the classroom. Enlarge and laminate if you wish. The term *denominator* will be very important in a few days when students need to find common denominators for addition of fractions. For now, they will have a little practice in finding denominators on a worksheet.

OPTIONAL: You might want to run off a Numerator-Denominator Wall Chart for each student to take home and study or have them make their own on an index card or piece of construction paper. If so, you should also send home Parents' Note 1 (page 11). You could require students to learn to spell both terms. They are easy to learn—both are spelled phonetically:

NU-MER-A-TOR

DE-NOM-I-NA-TOR

FOLDERS: Be sure each student buys or makes a folder to keep all *Math Phonics*™ materials organized. (See page 7.)

HOMEWORK: Assign Worksheet A (page 12). You may explain the Challenge section of the worksheet if you wish. However, students should be able to figure it out on their own. When you check Worksheet A, be sure to discuss what students have learned about how fractions are used by adults. Students are much more motivated to learn if they believe what they are learning is important.

NOTE: When checking Worksheet B (page 14), mention to students that 1. a. and 1. b. show $1/3$ and $2/6$. Though they are different fractions, they represent the same amount of a rectangle. They are called equivalent fractions, and we will work more with them later. Some math books refer to these fractions as **equal fractions**. If your students have been taught this as equal fractions, be sure they know both are correct.

OPTIONAL: If students need a little more time to understand the idea of a fraction, use Worksheet B. Challenge problems can be used as a regular part of the worksheet if needed. They are only slightly harder than the other problems.

Dear Parents,

We are beginning our unit on fractions, and we will need some support from you at home. We will use the *Math Phonics™—Fractions* system, and we will be sending home several study charts which should be posted at home. Please help your child find a good place for these charts where they will be seen several times every day. The bathroom mirror, the light switch in the bedroom or the side of the fridge might be some good spots.

Quiz your child verbally two or three times about the material on each chart until you think he or she has learned it.

Thanks for your help!

Sincerely,

Dear Parents,

In _____ days we will take an assessment or quiz about the basic addition and subtraction facts. Please review these with your child. Students must know these facts in order to learn about fractions!

Thanks for your help!

Sincerely,

WHAT IS A FRACTION?

1. Shade in the fraction that is named.

a. 1/4   b. 2/3   c. 6/6

2. Circle the numerator. Draw an X on the denominator.   2/3

3. Circle the fractions with the same denominator.

a. 1/2, 1/4, 3/4   b. 9/10, 2/10, 2/5   c. 2/3, 2/5, 1/3   d. 1/5, 3/5, 3/4

4. Three of these seven marbles are black. We say 3/7 are black. What fraction tells how many marbles are white? _____

5. Write the fraction for

a. black stars _____

b. white stars _____

c. gray stars _____

6. On the back, write at least five ways that adults use fractions at work or home. Ask your parents or other adults for ideas.

CHALLENGE:

1. Look at the stars in 5. Write the fraction for

   a. stars that are not white _____

   b. stars that are not black _____

   c. stars that are not gray _____

2. Your mom says that you may have 1/6 of the pizza. Draw a picture on the back of 1/6 of a pizza.

3. You have 1/3 of a giant-sized candy bar. Four people want to share that third. What fractional part of the whole candy bar would each person get? _____ Draw a picture on the back to help solve this.

NUMERATOR-DENOMINATOR

3 = AREAS SHADED = NUMERATOR

4 = PARTS IN ALL = DENOMINATOR

NU-MER-A-TOR

DE-NOM-I-NA-TOR

WHAT IS A FRACTION?

1. Shade in the correct part.

   a. $1/3$ b. $2/6$ c. $3/4$ d. $6/8$

2. Circle each fraction that has 5 as the denominator.

   $1/5$, $5/6$, $5/8$, $3/5$, $5/5$, $5/9$, $5/10$, $4/5$

3. Write the fraction for each.

   a. five-sixths _____   b. three-fourths _____

   c. nine-tenths _____   d. one-fifth _____

4. Write the fraction for

   a. large buttons _____   b. small buttons _____

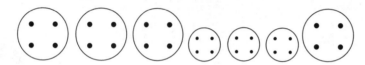

5. Write the words for the numbers above and below the line in a fraction. Ask your teacher to listen to you pronounce each word. _____   _____

CHALLENGE:

1. How many students are in your class? _____ How many girls? _____

   How many boys? _____

   Write a fraction for girls in the class. _____

   Write a fraction for boys in the class. _____

2. Write a fraction for the number of days you are usually in school in one week over the number of days in one week. _____

3. Fifteen students want ice cream. Five want chocolate, six want vanilla and four want strawberry. Write a fraction for students who want strawberry. _____

   Write a fraction for students who want chocolate. _____

LESSON PLAN 2

INTRODUCTION: In this lesson, we will cover the very simplest form of addition and subtraction of fractions. There will be no need to change denominators, no carrying, no borrowing, no improper fractions, no mixed numbers and no reducing.

The purpose of this lesson is to impress upon the students that when adding or subtracting with the same denominators, you add or subtract numerators—denominators stay the same.

ASSESSMENT: If your class has not been adding and subtracting recently, give the Addition and Subtraction Assessments (pages 19 and 21) to them. Give students and their parents two or three days' notice so they can review the addition and subtraction facts for the assessments. Send home Parents' Note 2 (page 11). You can also go over the page of addition and subtraction rules (page 17) with your class.

This is a very important step. If your students do not know their basic math facts, they will do poorly with fractions! They may understand the ideas involved in fractions quite well, but they will not get the problems right unless they can add and subtract! If you have a large number of students who do not know these facts, take a few days and go over the *Math Phonics™—Addition* and *Subtraction* books.

Many people go through life adding and subtracting on their fingers. They may do this mentally, but they are still counting out answers. This may work for some math problems, but in fractions it gets very confusing because one problem could have 10 or 12 individual steps. The student who is counting out answers gets lost, makes mistakes and gets discouraged. Students need to *memorize* answers to the math facts. *Math Phonics™* workbooks have excellent ways of doing this.

BASIC ADDITION AND SUBTRACTION OF FRACTIONS:
As often as possible, present new ideas with real-life examples. Use apples, pizzas or cherry pies for circles. Use boards, ribbons, yards of fabric or candy bars for rectangles. You could give each student a set of fraction circles included in Lesson Plan 7 or a set of fraction strips included in Lesson Plan 3 if you wish.

Example: This can be demonstrated with a drawing on the board or with a cardboard circle cut into thirds. Dan has $1/3$ of his pizza to warm up in the microwave. His sister Amanda has $1/3$ of her pizza to warm up, also. If they put both pieces together on a plate, how much of the pizza is on that plate?

$$1/3 + 1/3 = 2/3$$

Use another example with a cardboard rectangle or a drawing of a rectangle on the board. Rachel cut her banana bread into five pieces. She ate $1/5$ and gave $1/5$ to Hannah to eat. How much of the bread have the two girls eaten?

$$1/5 + 1/5 = 2/5$$

Now ask your class if they can make up a rule to help them add fractions.

RULE: *Addition of Fractions.* Add the two top numbers (numerators). Keep the same bottom number (denominator).

Now try subtraction. Zach has $4/5$ of a candy bar. He gives away $1/5$ of it. Now how much is left?

$$4/5 - 1/5 = 3/5$$

What could be the rule for subtraction?

RULE: *Subtraction of Fractions.* Subtract the two numerators. Keep the same denominator.

Try some harder problems like $15/20 - 7/20$ or $23/25 - 16/25$. Tell students to use scratch paper to subtract numerators if they need to.

Show both forms for writing fraction problems.

$$1/3 + 1/3 = 2/3 \quad \text{or} \quad \begin{array}{r} \frac{1}{3} \\ + \frac{1}{3} \\ \hline \frac{2}{3} \end{array}$$

Show the same for subtraction.

OPTIONAL: Enlarge and laminate Addition and Subtraction of Fractions Wall Chart B (page 22) with rules. Send home a small chart with each student to study, or have students make their own on index cards or construction paper.

NOTE: Starting with this lesson, there will be three worksheets for each lesson. The third worksheet, called Level 2, can be used for advanced students, or can be used when students review these concepts. When going over fractions later, you could use one of the first two worksheets as a review and the Level 2 worksheet as well.

ADDITION RULES

EVENS AND ODDS: An even number always ends in 0, 2, 4, 6 or 8.

An odd number always ends in 1, 3, 5, 7 or 9.

0s: When you add zero to a number, the number stays the same.

1s: When you add one to a number, the answer is the next number on the number line.

2s: When adding two to an even number, the answer is the next even number. (4 + 2 = 6, 6 + 2 = 8)

When adding two to an odd number, the answer is the next odd number. (3 + 2 = 5, 5 + 2 = 7)

DOUBLES: Answers to the doubles (2 + 2, 3 + 3, etc.) are always even.

NUMBER NEIGHBORS: Number neighbors are any two numbers side-by-side on the number line. (Ex: 5 and 6 or 6 and 7)

When you add two number neighbors, the answer is always an odd number. (Ex: 5 + 6 = 11, 6 + 7 = 13). To add two number neighbors, double the smaller number and add 1. (Since 5 + 5 = 10, you know 5 + 6 = 11.)

Whenever you add an even and an odd number, the answer is an odd number. (3 + 6 = 9, 7 + 8 = 15)

9s: To add nine to a number, first subtract one from the number. Then put a 1 in front of that answer. (6 + 9 = 15, 8 + 9 = 17)

SUBTRACTION RULES

NUMBER NEIGHBORS: When you subtract a smaller number neighbor from a larger number neighbor, the answer is always one. (8 - 7 = 1, 9 - 8 = 1)

When you subtract a larger number neighbor from a smaller number neighbor and you can borrow, the answer is nine. (17 - 8 = 9, 15 - 6 = 9)

1s: When you subtract one from a number, the answer is the smaller number neighbor.

9s: When you subtract nine from a teens number, add the numerals of the teens number. That is the answer. (15 - 9 = 6, 13 - 9 = 4)

2s: When you subtract two from an odd number, the answer is the next smaller odd number. (7 - 2 = 5, 9 - 2 = 7)

When you subtract two from an even number, the answer is the next smaller even number. (8 - 2 = 6, 6 - 2 = 4)

When you subtract two closest odd numbers, the answer is two. (7 - 5 = 2, 9 - 7 = 2)

When you subtract two closest even numbers, the answer is two. (8 - 6 = 2, 6 - 4 = 2)

ADDITION FACTS

Zeros	1s	2s	3s	4s
0 + 0 = 0	1 + 1 = 2	2 + 2 = 4	3 + 3 = 6	4 + 4 = 8
0 + 1 = 1	1 + 2 = 3	2 + 4 = 6	3 + 4 = 7	4 + 5 = 9
0 + 2 = 2	1 + 3 = 4	2 + 6 = 8	3 + 5 = 8	4 + 6 = 10
0 + 3 = 3	1 + 4 = 5	2 + 8 = 10	3 + 6 = 9	4 + 7 = 11
0 + 4 = 4	1 + 5 = 6		3 + 7 = 10	4 + 8 = 12
0 + 5 = 5	1 + 6 = 7	2 + 3 = 5	3 + 8 = 11	4 + 9 = 13
0 + 6 = 6	1 + 7 = 8	2 + 5 = 7	3 + 9 = 12	
0 + 7 = 7	1 + 8 = 9	2 + 7 = 9		
0 + 8 = 8	1 + 9 = 10	2 + 9 = 11		
0 + 9 = 9				

5s	6s	7s	8s	9s
5 + 5 = 10	6 + 6 = 12	7 + 7 = 14	8 + 8 = 16	9 + 9 = 18
5 + 6 = 11	6 + 7 = 13	7 + 8 = 15	8 + 9 = 17	
5 + 7 = 12	6 + 8 = 14	7 + 9 = 16		
5 + 8 = 13	6 + 9 = 15			
5 + 9 = 14				

18

Name _____

ADDITION ASSESSMENT

1. **5 + 4** = _____   2. **3 + 1** = _____   3. **10 + 10** = _____

4. **2 + 0** = _____   5. **5 + 5** = _____   6. **3 + 3** = _____

7. **3 + 2** = _____   8. **6 + 6** = _____   9. **0 + 0** = _____

10. **9 + 4** = _____   11. **1 + 2** = _____   12. **5 + 2** = _____

13. **4 + 0** = _____   14. **1 + 8** = _____   15. **8 + 8** = _____

16. **4 + 3** = _____   17. **9 + 1** = _____   18. **1 + 4** = _____

19. **6 + 3** = _____   20. **0 + 1** = _____   21. **10 + 4** = _____

22. **6 + 5** = _____   23. **0 + 3** = _____   24. **7 + 7** = _____

25. **9 + 8** = _____   26. **8 + 7** = _____   27. **1 + 1** = _____

28. **3 + 7** = _____   29. **5 + 1** = _____   30. **5 + 7** = _____

31. **10 + 3** = _____   32. **4 + 4** = _____   33. **7 + 9** = _____

34. **0 + 5** = _____   35. **8 + 3** = _____   36. **4 + 6** = _____

37. **1 + 6** = _____   38. **6 + 0** = _____   39. **3 + 9** = _____

40. **2 + 6** = _____   41. **7 + 1** = _____   42. **9 + 2** = _____

43. **0 + 7** = _____   44. **2 + 2** = _____   45. **7 + 6** = _____

46. **10 + 8** = _____   47. **2 + 8** = _____   48. **8 + 5** = _____

49. **7 + 2** = _____   50. **7 + 4** = _____   51. **10 + 1** = _____

52. **8 + 0** = _____   53. **5 + 9** = _____   54. **9 + 6** = _____

55. **2 + 4** = _____   56. **0 + 9** = _____   57. **3 + 5** = _____

58. **10 + 5** = _____   59. **10 + 0** = _____   60. **10 + 9** = _____

61. **6 + 8** = _____   62. **4 + 8** = _____   63. **10 + 7** = _____

64. **10 + 6** = _____   65. **10 + 2** = _____   66. **9 + 9** = _____

SUBTRACTION FACTS

0s	1s	2s	3s	4s
0 - 0 = 0	1 - 1 = 0	2 - 2 = 0	3 - 3 = 0	4 - 4 = 0
	1 - 0 = 1	2 - 1 = 1	3 - 2 = 1	4 - 3 = 1
		2 - 0 = 2	3 - 1 = 2	4 - 2 = 2
			3 - 0 = 3	4 - 1 = 3
				4 - 0 = 4

5s	6s	7s	8s	9s
5 - 5 = 0	6 - 6 = 0	7 - 7 = 0	8 - 8 = 0	9 - 9 = 0
5 - 4 = 1	6 - 5 = 1	7 - 6 = 1	8 - 7 = 1	9 - 8 = 1
5 - 3 = 2	6 - 4 = 2	7 - 5 = 2	8 - 6 = 2	9 - 7 = 2
5 - 2 = 3	6 - 3 = 3	7 - 4 = 3	8 - 5 = 3	9 - 6 = 3
5 - 1 = 4	6 - 2 = 4	7 - 3 = 4	8 - 4 = 4	9 - 5 = 4
5 - 0 = 5	6 - 1 = 5	7 - 2 = 5	8 - 3 = 5	9 - 4 = 5
	6 - 0 = 6	7 - 1 = 6	8 - 2 = 6	9 - 3 = 6
		7 - 0 = 7	8 - 1 = 7	9 - 2 = 7
			8 - 0 = 8	9 - 1 = 8
				9 - 0 = 9

10s	11s	12s	13s	14s
10 - 9 = 1	11 - 9 = 2	12 - 9 = 3	13 - 9 = 4	14 - 9 = 5
10 - 8 = 2	11 - 8 = 3	12 - 8 = 4	13 - 8 = 5	14 - 8 = 6
10 - 7 = 3	11 - 7 = 4	12 - 7 = 5	13 - 7 = 6	14 - 7 = 7
10 - 6 = 4	11 - 6 = 5	12 - 6 = 6	13 - 6 = 7	14 - 6 = 8
10 - 5 = 5	11 - 5 = 6	12 - 5 = 7	13 - 5 = 8	14 - 5 = 9
10 - 4 = 6	11 - 4 = 7	12 - 4 = 8	13 - 4 = 9	
10 - 3 = 7	11 - 3 = 8	12 - 3 = 9		
10 - 2 = 8	11 - 2 = 9			
10 - 1 = 9				

15s	16s	17s	18s
15 - 9 = 6	16 - 9 = 7	17 - 9 = 8	18 - 9 = 9
15 - 8 = 7	16 - 8 = 8	17 - 8 = 9	
15 - 7 = 8	16 - 7 = 9		
15 - 6 = 9			

Notice that starting with the 10s, we do not include 10 - 10 = 0, 11 - 11 = 0, 11 - 10 = 1 and so on. The reason is that students have already had 1 - 1 = 0 and 0 - 0 = 0, so there is no need to repeat those with the 10s, 11s and larger numbers.

SUBTRACTION ASSESSMENT

1. $\begin{array}{r} 7 \\ -6 \\ \hline \end{array}$ 2. $\begin{array}{r} 11 \\ -2 \\ \hline \end{array}$ 3. $\begin{array}{r} 4 \\ -3 \\ \hline \end{array}$ 4. $\begin{array}{r} 8 \\ -7 \\ \hline \end{array}$ 5. $\begin{array}{r} 9 \\ -4 \\ \hline \end{array}$ 6. $\begin{array}{r} 5 \\ -0 \\ \hline \end{array}$ 7. $\begin{array}{r} 10 \\ -10 \\ \hline \end{array}$ 8. $\begin{array}{r} 16 \\ -8 \\ \hline \end{array}$ 9. $\begin{array}{r} 11 \\ -8 \\ \hline \end{array}$

10. $\begin{array}{r} 12 \\ -6 \\ \hline \end{array}$ 11. $\begin{array}{r} 10 \\ -9 \\ \hline \end{array}$ 12. $\begin{array}{r} 14 \\ -5 \\ \hline \end{array}$ 13. $\begin{array}{r} 11 \\ -5 \\ \hline \end{array}$ 14. $\begin{array}{r} 10 \\ -8 \\ \hline \end{array}$ 15. $\begin{array}{r} 15 \\ -7 \\ \hline \end{array}$ 16. $\begin{array}{r} 10 \\ -7 \\ \hline \end{array}$ 17. $\begin{array}{r} 8 \\ -8 \\ \hline \end{array}$ 18. $\begin{array}{r} 7 \\ -7 \\ \hline \end{array}$

19. $\begin{array}{r} 8 \\ -4 \\ \hline \end{array}$ 20. $\begin{array}{r} 8 \\ -1 \\ \hline \end{array}$ 21. $\begin{array}{r} 13 \\ -6 \\ \hline \end{array}$ 22. $\begin{array}{r} 11 \\ -6 \\ \hline \end{array}$ 23. $\begin{array}{r} 7 \\ -5 \\ \hline \end{array}$ 24. $\begin{array}{r} 11 \\ -7 \\ \hline \end{array}$ 25. $\begin{array}{r} 2 \\ -0 \\ \hline \end{array}$ 26. $\begin{array}{r} 9 \\ -3 \\ \hline \end{array}$ 27. $\begin{array}{r} 4 \\ -1 \\ \hline \end{array}$

28. $\begin{array}{r} 1 \\ -1 \\ \hline \end{array}$ 29. $\begin{array}{r} 12 \\ -8 \\ \hline \end{array}$ 30. $\begin{array}{r} 16 \\ -9 \\ \hline \end{array}$ 31. $\begin{array}{r} 13 \\ -8 \\ \hline \end{array}$ 32. $\begin{array}{r} 6 \\ -6 \\ \hline \end{array}$ 33. $\begin{array}{r} 3 \\ -0 \\ \hline \end{array}$ 34. $\begin{array}{r} 15 \\ -9 \\ \hline \end{array}$ 35. $\begin{array}{r} 13 \\ -7 \\ \hline \end{array}$ 36. $\begin{array}{r} 6 \\ -1 \\ \hline \end{array}$

37. $\begin{array}{r} 12 \\ -3 \\ \hline \end{array}$ 38. $\begin{array}{r} 5 \\ -4 \\ \hline \end{array}$ 39. $\begin{array}{r} 14 \\ -7 \\ \hline \end{array}$ 40. $\begin{array}{r} 6 \\ -3 \\ \hline \end{array}$ 41. $\begin{array}{r} 10 \\ -6 \\ \hline \end{array}$ 42. $\begin{array}{r} 9 \\ -0 \\ \hline \end{array}$ 43. $\begin{array}{r} 14 \\ -8 \\ \hline \end{array}$ 44. $\begin{array}{r} 13 \\ -9 \\ \hline \end{array}$ 45. $\begin{array}{r} 8 \\ -0 \\ \hline \end{array}$

46. $\begin{array}{r} 12 \\ -9 \\ \hline \end{array}$ 47. $\begin{array}{r} 9 \\ -9 \\ \hline \end{array}$ 48. $\begin{array}{r} 5 \\ -2 \\ \hline \end{array}$ 49. $\begin{array}{r} 11 \\ -7 \\ \hline \end{array}$ 50. $\begin{array}{r} 11 \\ -3 \\ \hline \end{array}$ 51. $\begin{array}{r} 15 \\ -6 \\ \hline \end{array}$ 52. $\begin{array}{r} 9 \\ -8 \\ \hline \end{array}$ 53. $\begin{array}{r} 11 \\ -9 \\ \hline \end{array}$ 54. $\begin{array}{r} 10 \\ -2 \\ \hline \end{array}$

55. $\begin{array}{r} 5 \\ -3 \\ \hline \end{array}$ 56. $\begin{array}{r} 5 \\ -5 \\ \hline \end{array}$ 57. $\begin{array}{r} 3 \\ -3 \\ \hline \end{array}$ 58. $\begin{array}{r} 16 \\ -7 \\ \hline \end{array}$ 59. $\begin{array}{r} 9 \\ -1 \\ \hline \end{array}$ 60. $\begin{array}{r} 12 \\ -7 \\ \hline \end{array}$ 61. $\begin{array}{r} 6 \\ -0 \\ \hline \end{array}$ 62. $\begin{array}{r} 4 \\ -0 \\ \hline \end{array}$ 63. $\begin{array}{r} 10 \\ -1 \\ \hline \end{array}$

64. $\begin{array}{r} 5 \\ -1 \\ \hline \end{array}$ 65. $\begin{array}{r} 7 \\ -1 \\ \hline \end{array}$ 66. $\begin{array}{r} 10 \\ -4 \\ \hline \end{array}$ 67. $\begin{array}{r} 15 \\ -8 \\ \hline \end{array}$ 68. $\begin{array}{r} 12 \\ -1 \\ \hline \end{array}$ 69. $\begin{array}{r} 9 \\ -6 \\ \hline \end{array}$ 70. $\begin{array}{r} 8 \\ -3 \\ \hline \end{array}$ 71. $\begin{array}{r} 2 \\ -1 \\ \hline \end{array}$ 72. $\begin{array}{r} 3 \\ -2 \\ \hline \end{array}$

73. $\begin{array}{r} 14 \\ -9 \\ \hline \end{array}$ 74. $\begin{array}{r} 18 \\ -9 \\ \hline \end{array}$ 75. $\begin{array}{r} 17 \\ -8 \\ \hline \end{array}$ 76. $\begin{array}{r} 6 \\ -4 \\ \hline \end{array}$ 77. $\begin{array}{r} 0 \\ -0 \\ \hline \end{array}$ 78. $\begin{array}{r} 7 \\ -4 \\ \hline \end{array}$ 79. $\begin{array}{r} 6 \\ -2 \\ \hline \end{array}$ 80. $\begin{array}{r} 10 \\ -3 \\ \hline \end{array}$ 81. $\begin{array}{r} 7 \\ -0 \\ \hline \end{array}$

82. $\begin{array}{r} 9 \\ -8 \\ \hline \end{array}$ 83. $\begin{array}{r} 13 \\ -4 \\ \hline \end{array}$ 84. $\begin{array}{r} 9 \\ -4 \\ \hline \end{array}$ 85. $\begin{array}{r} 17 \\ -9 \\ \hline \end{array}$ 86. $\begin{array}{r} 10 \\ -5 \\ \hline \end{array}$ 87. $\begin{array}{r} 6 \\ -5 \\ \hline \end{array}$ 88. $\begin{array}{r} 12 \\ -4 \\ \hline \end{array}$ 89. $\begin{array}{r} 1 \\ -0 \\ \hline \end{array}$ 90. $\begin{array}{r} 7 \\ -3 \\ \hline \end{array}$

91. $\begin{array}{r} 9 \\ -7 \\ \hline \end{array}$ 92. $\begin{array}{r} 2 \\ -2 \\ \hline \end{array}$ 93. $\begin{array}{r} 14 \\ -6 \\ \hline \end{array}$ 94. $\begin{array}{r} 13 \\ -5 \\ \hline \end{array}$ 95. $\begin{array}{r} 8 \\ -5 \\ \hline \end{array}$ 96. $\begin{array}{r} 3 \\ -1 \\ \hline \end{array}$ 97. $\begin{array}{r} 4 \\ -4 \\ \hline \end{array}$ 98. $\begin{array}{r} 8 \\ -2 \\ \hline \end{array}$ 99. $\begin{array}{r} 4 \\ -2 \\ \hline \end{array}$

100. $\begin{array}{r} 8 \\ -6 \\ \hline \end{array}$ 101. $\begin{array}{r} 7 \\ -2 \\ \hline \end{array}$

ADDITION AND SUBTRACTION OF FRACTIONS

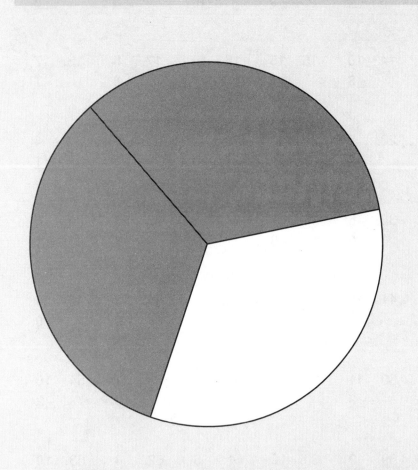

$$\frac{1}{3} + \frac{1}{3} = \frac{2}{3}$$

ADD NUMERATORS
KEEP SAME DENOMINATOR

$$\frac{4}{5} - \frac{1}{5} = \frac{3}{5}$$

SUBTRACT NUMERATORS
KEEP SAME DENOMINATOR

SIMPLE ADDITION AND SUBTRACTION

Directions: Add numerators. Keep same denominators.

1. $\frac{1}{4} + \frac{1}{4} =$ ____

2. $\frac{2}{5} + \frac{2}{5} =$ ____

3. $\frac{5}{11} + \frac{3}{11} =$ ____

4. $\frac{3}{10} + \frac{4}{10} =$ ____

5. $\frac{7}{20} + \frac{9}{20} =$ ____

6. $\frac{7}{15} + \frac{7}{15} =$ ____

Directions: Subtract numerators. Keep the same denominators.

7. $\frac{3}{5}$
$-\ \frac{1}{5}$

8. $\frac{12}{15}$
$-\ \frac{6}{15}$

9. $\frac{15}{24}$
$-\ \frac{9}{24}$

10. $\frac{11}{20}$
$-\ \frac{6}{20}$

11. $\frac{14}{18}$
$-\ \frac{7}{18}$

12. $\frac{17}{21}$
$-\ \frac{12}{21}$

Directions: Add or subtract numerators. Watch the signs!

13. $\frac{9}{18}$
$+\ \frac{6}{18}$

14. $\frac{2}{13}$
$+\ \frac{8}{13}$

15. $\frac{16}{22}$
$-\ \frac{7}{22}$

16. $\frac{11}{13}$
$-\ \frac{4}{13}$

17. $\frac{19}{20}$
$-\ \frac{8}{20}$

18. $\frac{5}{14}$
$+\ \frac{8}{14}$

19. You live $3/10$ of a mile from school. It is another $4/10$ of a mile to the library. How far is it from your house to the library? _____

20. You and your brother each ate $2/5$ of a pizza for lunch. How much is left? _____

21. You have $1/6$ of the money you need to buy a watch. If you earn another $3/6$ of the money, how much do you have in all? _____

CHALLENGE:

You have three dimes in your pocket. What fraction of a dollar do you have in your pocket? _____

What is another way of writing about three dimes without using a fraction? _____

ADDITION AND SUBTRACTION OF FRACTIONS

Directions: Add or subtract numerators. Watch the signs!

1. $\frac{2}{7}$ 2. $\frac{3}{8}$ 3. $\frac{1}{10}$ 4. $\frac{7}{15}$ 5. $\frac{7}{17}$ 6. $\frac{7}{13}$
   $+\frac{3}{7}$ $+\frac{2}{8}$ $+\frac{5}{10}$ $+\frac{6}{15}$ $+\frac{8}{17}$ $+\frac{5}{13}$

7. $\frac{9}{10}$ 8. $\frac{13}{15}$ 9. $\frac{7}{10}$ 10. $\frac{15}{16}$ 11. $\frac{11}{12}$ 12. $\frac{12}{13}$
   $-\frac{7}{10}$ $-\frac{8}{15}$ $-\frac{3}{10}$ $-\frac{8}{16}$ $-\frac{5}{12}$ $-\frac{7}{13}$

13. If you have $2/7$ of a gallon of juice and $4/7$ of a gallon of club soda and mix them together, how much punch will they make? _____

14. You have done $3/10$ of your homework at school and $4/10$ of it before supper at home. How much more do you have to do? _____

Find these words hidden in the word search below: addition, denominator, divide, fraction, multiply, numerator, product, reduce, subtract, sum. Words can be found across or down.

```
D  E  N  O  M  I  N  A  T  O  R
I  S  U  M  A  Y  U  N  S  E  E
V  T  M  P  N  A  M  D  U  M  D
I  A  R  R  I  D  E  R  B  U  U
D  T  T  O  L  D  R  A  T  L  C
E  E  O  D  R  I  A  S  R  T  E
A  A  R  U  S  T  T  N  A  I  L
F  R  A  C  T  I  O  N  C  P  I
N  E  S  T  G  O  R  O  T  L  O
G  D  P  A  I  N  U  T  S  Y  N
```

CHALLENGE:
Circle all the other words you can find in this word search. Write them on the back.

Name _____

SIMPLE ADDITION AND SUBTRACTION

Directions: Add numerators. Keep the same denominators.

1. $\frac{5}{20}$ $+\frac{8}{20}$

2. $\frac{13}{30}$ $+\frac{16}{30}$

3. $\frac{11}{40}$ $+\frac{22}{40}$

4. $\frac{23}{50}$ $+\frac{19}{50}$

5. $\frac{37}{60}$ $+\frac{14}{60}$

6. $\frac{29}{60}$ $+\frac{17}{60}$

Directions: Subtract numerators. Keep the same denominators.

7. $\frac{27}{30}$ $-\frac{19}{30}$

8. $\frac{35}{40}$ $-\frac{18}{40}$

9. $\frac{28}{30}$ $-\frac{19}{30}$

10. $\frac{37}{40}$ $-\frac{16}{40}$

11. $\frac{41}{50}$ $-\frac{38}{50}$

12. $\frac{25}{30}$ $-\frac{18}{30}$

Directions: Add or subtract numerators. Watch the signs!

13. $\frac{3}{20}$ $+\frac{14}{20}$

14. $\frac{19}{20}$ $-\frac{12}{20}$

15. $\frac{23}{30}$ $-\frac{9}{30}$

16. $\frac{24}{60}$ $+\frac{19}{60}$

17. $\frac{18}{50}$ $+\frac{13}{50}$

18. $\frac{42}{50}$ $-\frac{29}{50}$

19. Of the 35 people in the class, 15 are allergic to chocolate and 7 do not like chocolate. What fractional part of the class will probably not be eating chocolate? _____

20. $7/18$ of the students made the track team. $9/18$ of the students made the volleyball team. (No one made both teams.) What fractional part of the students are on those teams? _____

21. $9/10$ of the class tried out for the school play. $3/10$ of the class were picked. What fraction of the class tried out and were not picked? _____

CHALLENGE:
Here is a magic square. In every row of three fractions–down, across or diagonal–the sum is the same. Fill in the fractions.

$\frac{6}{15}$		$\frac{8}{15}$
	$\frac{5}{15}$	
$\frac{2}{15}$		

LESSON PLAN 3

INTRODUCTION: In this section we will focus on changing from one equivalent fraction to another. This was touched upon in Worksheet B (page 14).

MULTIPLICATION ASSESSMENT: In order to change from one fraction to an equivalent one, students will be multiplying the numerator and denominator by the same number. If your class has not been using multiplication recently, give students the Multiplication Assessment (pages 32-33). Send home Parents' Note 3 (page 29) several days before the assessment. Give students the multiplication rules included. If you have several students who do not know their multiplication facts, take a few days and review all the facts using the *Math Phonics™–Multiplication* book.

Students will not be able to work with fractions without a thorough knowledge of multiplication facts.

EQUIVALENT FRACTIONS: Remind students of Worksheet B in which they shaded rectangles for $1/3$ and $2/6$. Draw rectangles on the board to demonstrate these two fractions.

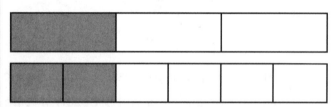

$1/3$ and $2/6$ are equal in value, but they are written using different numbers. That's why we call them equivalent fractions. We don't want to have to draw rectangles all the time. Another way to find equivalent fractions is to multiply the numerator and denominator by the same number.

$$\frac{1}{3} \begin{array}{c} \times 2 \\ \times 2 \end{array} = \frac{2}{6}$$

We will demonstrate a few more by use of rectangles so students will believe that multiplying really works. Show them $3/4 = 6/8$ and $5/10 = 1/2$.

(When we multiply the numerator and denominator by the same number like $2/2$, we are really multiplying by one. We will talk about this more in the lesson on multiplying fractions.)

EQUIVALENT FRACTION STRIPS: Give each student a page of Equivalent Fraction Strips (page 28) and an 8½" x 4" envelope for storing the strips in a pocket of their Math Phonics™ folders. Laminating the pages is a good idea, but not essential. Be sure to tell them to use them carefully so they will last until the class is finished with fractions.

Make your own set of Equivalent Fraction Strips for the overhead projector or enlarge and laminate for use as a wall chart.

Ask the class to find another fraction on the fraction strips which is equivalent to ¼. Students should find ²/₈ and ³/₁₂.

Write these pairs on the board. ¼ = ²/₈. What number was used to multiply?

The answer is 2. What was multiplied to change from ¼ to ³/₁₂? The answer is 3.

$$\frac{1 \ (\times 3)}{4 \ (\times 3)} = \frac{3}{12}$$

Could they change from ¼ to ⅙? No. There is no number times 4 that equals 6. (Someone might say multiply by 1½. That is correct, but right now we are multiplying by whole numbers.)

STUDY BUDDIES: Divide the class into math study buddy groups of three or four students each. Have them see how many different equivalent pairs of fractions they can find using the Equivalent Fraction Strips. Each student should write the pairs of equivalent fractions on a sheet of paper to hand in.

MULTIPLES: Give each student a sheet of multiples 2 through 12 (page 34). Have the class chant the multiples for a few days until students are able to recognize multiples easily.

WORKSHEET G (PAGE 36): Show students the greater than (>), less than (<) and equal (=) signs. Have students use their Equivalent Fraction Strips to find out which fraction is larger.

½ and ⅓,   ¼ and ⅕.   Think of some other pairs.

To simplify the less than and greater than signs, tell students to have the large part of the signs point toward the larger fraction.

1											

| $\frac{1}{2}$ | | | | | | $\frac{1}{2}$ | | | | | |

| $\frac{1}{3}$ | | | | $\frac{1}{3}$ | | | | $\frac{1}{3}$ | | | |

| $\frac{1}{4}$ | | | $\frac{1}{4}$ | | | $\frac{1}{4}$ | | | $\frac{1}{4}$ | | |

| $\frac{1}{5}$ | | $\frac{1}{5}$ | | $\frac{1}{5}$ | | $\frac{1}{5}$ | | $\frac{1}{5}$ | | | |

| $\frac{1}{6}$ | | $\frac{1}{6}$ | | $\frac{1}{6}$ | | $\frac{1}{6}$ | | $\frac{1}{6}$ | | $\frac{1}{6}$ | |

| $\frac{1}{8}$ | $\frac{1}{8}$ | $\frac{1}{8}$ | $\frac{1}{8}$ | $\frac{1}{8}$ | $\frac{1}{8}$ | $\frac{1}{8}$ | $\frac{1}{8}$ | | | | |

| $\frac{1}{10}$ | $\frac{1}{10}$ | $\frac{1}{10}$ | $\frac{1}{10}$ | $\frac{1}{10}$ | $\frac{1}{10}$ | $\frac{1}{10}$ | $\frac{1}{10}$ | $\frac{1}{10}$ | $\frac{1}{10}$ | | |

| $\frac{1}{12}$ | $\frac{1}{12}$ | $\frac{1}{12}$ | $\frac{1}{12}$ | $\frac{1}{12}$ | $\frac{1}{12}$ | $\frac{1}{12}$ | $\frac{1}{12}$ | $\frac{1}{12}$ | $\frac{1}{12}$ | $\frac{1}{12}$ | $\frac{1}{12}$ |

Dear Parents,

The class will have a review quiz of multiplication facts on _____.
Please help your child study the facts which are on the study sheet.

Go over the study sheets and wall charts which have been sent home in the past. Students need to remember what has been learned earlier because all the information will be used in the study of fractions.

Students also need to memorize the page of multiples so that they can list each group easily.

Thanks for your help!

Sincerely,

Dear Parents,

We will have an assessment of division facts on _____. Please help your child study these facts before the assessment.

Once again, thanks for working with your child at home. The one-on-one math study is extremely valuable for your child.

Sincerely,

P.S. Our class has learned how to play Fraction Slapjack. Ask your child how to play the game. Although this is a game, it is a very good way for students to practice reducing fractions. I hope you will play this and other math games several times a week. Students do best in math when they have several different ways of learning.

MULTIPLICATION FACTS

0 x 1 = 0		0 x 2 = 0		0 x 3 = 0		0 x 4 = 0
1 x 1 = 1		1 x 2 = 2		1 x 3 = 3		1 x 4 = 4
2 x 1 = 2		2 x 2 = 4		2 x 3 = 6		2 x 4 = 8
3 x 1 = 3		3 x 2 = 6		3 x 3 = 9		3 x 4 = 12
4 x 1 = 4		4 x 2 = 8		4 x 3 = 12		4 x 4 = 16
5 x 1 = 5		5 x 2 = 10		5 x 3 = 15		5 x 4 = 20
6 x 1 = 6		6 x 2 = 12		6 x 3 = 18		6 x 4 = 24
7 x 1 = 7		7 x 2 = 14		7 x 3 = 21		7 x 4 = 28
8 x 1 = 8		8 x 2 = 16		8 x 3 = 24		8 x 4 = 32
9 x 1 = 9		9 x 2 = 18		9 x 3 = 27		9 x 4 = 36
10 x 1 = 10		10 x 2 = 20		10 x 3 = 30		10 x 4 = 40

0 x 5 = 0		0 x 6 = 0		0 x 7 = 0
1 x 5 = 5		1 x 6 = 6		1 x 7 = 7
2 x 5 = 10		2 x 6 = 12		2 x 7 = 14
3 x 5 = 15		3 x 6 = 18		3 x 7 = 21
4 x 5 = 20		4 x 6 = 24		4 x 7 = 28
5 x 5 = 25		5 x 6 = 30		5 x 7 = 35
6 x 5 = 30		6 x 6 = 36		6 x 7 = 42
7 x 5 = 35		7 x 6 = 42		7 x 7 = 49
8 x 5 = 40		8 x 6 = 48		8 x 7 = 56
9 x 5 = 45		9 x 6 = 54		9 x 7 = 63
10 x 5 = 50		10 x 6 = 60		10 x 7 = 70

0 x 8 = 0		0 x 9 = 0		0 x 10 = 0
1 x 8 = 8		1 x 9 = 9		1 x 10 = 10
2 x 8 = 16		2 x 9 = 18		2 x 10 = 20
3 x 8 = 24		3 x 9 = 27		3 x 10 = 30
4 x 8 = 32		4 x 9 = 36		4 x 10 = 40
5 x 8 = 40		5 x 9 = 45		5 x 10 = 50
6 x 8 = 48		6 x 9 = 54		6 x 10 = 60
7 x 8 = 56		7 x 9 = 63		7 x 10 = 70
8 x 8 = 64		8 x 9 = 72		8 x 10 = 80
9 x 8 = 72		9 x 9 = 81		9 x 10 = 90
10 x 8 = 80		10 x 9 = 90		10 x 10 = 100

MULTIPLICATION FACTS

Take time to memorize the squares as a group.

```
 0 x  0 =   0
 1 x  1 =   1
 2 x  2 =   4
 3 x  3 =   9
 4 x  4 =  16
 5 x  5 =  25
 6 x  6 =  36
 7 x  7 =  49
 8 x  8 =  64
 9 x  9 =  81
10 x 10 = 100
```

The squares are extremely important in algebra and trigonometry. Some teachers even require students to memorize the squares up to 25 x 25 = 625 to help in solving quadratic equations and trigonometry problems.

Also, notice these patterns in the 9s.

```
 0 x 9 =  0
 1 x 9 =  9
 2 x 9 = 18
 3 x 9 = 27
 4 x 9 = 36
 5 x 9 = 45
 6 x 9 = 54
 7 x 9 = 63
 8 x 9 = 72
 9 x 9 = 81
10 x 9 = 90
```

For the answers 18 through 90, adding the two numerals gives nine. The number in the 10s place is one less than the number you are multiplying by nine.

With each answer, the numeral in the 10s place increases and the numeral in the 1s place decreases.

So if you are stuck trying to think of 6 x 9, remember that it is in the 50s because 5 is one less than 6. Since 5 + 4 = 9, the answer is 54.

MULTIPLICATION ASSESSMENT

1. $12 \times 2 =$ ___   2. $11 \times 6 =$ ___   3. $0 \times 1 =$ ___

4. $3 \times 1 =$ ___   5. $12 \times 3 =$ ___   6. $2 \times 2 =$ ___

7. $11 \times 3 =$ ___   8. $0 \times 0 =$ ___   9. $7 \times 5 =$ ___

10. $6 \times 9 =$ ___   11. $5 \times 6 =$ ___   12. $10 \times 4 =$ ___

13. $8 \times 7 =$ ___   14. $7 \times 7 =$ ___   15. $1 \times 2 =$ ___

16. $2 \times 5 =$ ___   17. $3 \times 4 =$ ___   18. $4 \times 0 =$ ___

19. $0 \times 2 =$ ___   20. $1 \times 6 =$ ___   21. $9 \times 4 =$ ___

22. $10 \times 6 =$ ___   23. $5 \times 3 =$ ___   24. $10 \times 7 =$ ___

25. $8 \times 8 =$ ___   26. $7 \times 4 =$ ___   27. $5 \times 8 =$ ___

28. $4 \times 6 =$ ___   29. $12 \times 6 =$ ___   30. $1 \times 1 =$ ___

31. $9 \times 9 =$ ___   32. $10 \times 11 =$ ___   33. $2 \times 6 =$ ___

34. $4 \times 2 =$ ___   35. $3 \times 3 =$ ___   36. $11 \times 8 =$ ___

37. $7 \times 9 =$ ___   38. $3 \times 0 =$ ___   39. $11 \times 11 =$ ___

40. $9 \times 8 =$ ___   41. $9 \times 5 =$ ___   42. $2 \times 3 =$ ___

43. $10 \times 10 =$ ___   44. $4 \times 8 =$ ___   45. $10 \times 5 =$ ___

MULTIPLICATION ASSESSMENT

46. $11 \times 6 =$ ___   47. $8 \times 2 =$ ___   48. $3 \times 8 =$ ___

49. $0 \times 5 =$ ___   50. $11 \times 2 =$ ___   51. $12 \times 0 =$ ___

52. $6 \times 3 =$ ___   53. $0 \times 8 =$ ___   54. $11 \times 2 =$ ___

55. $11 \times 9 =$ ___   56. $10 \times 3 =$ ___   57. $4 \times 4 =$ ___

58. $1 \times 4 =$ ___   59. $9 \times 1 =$ ___   60. $11 \times 1 =$ ___

61. $11 \times 5 =$ ___   62. $2 \times 9 =$ ___   63. $5 \times 1 =$ ___

64. $12 \times 8 =$ ___   65. $6 \times 8 =$ ___   66. $11 \times 0 =$ ___

67. $10 \times 1 =$ ___   68. $10 \times 8 =$ ___   69. $9 \times 3 =$ ___

70. $10 \times 9 =$ ___   71. $10 \times 12 =$ ___   72. $5 \times 5 =$ ___

73. $2 \times 7 =$ ___   74. $1 \times 8 =$ ___   75. $12 \times 9 =$ ___

76. $12 \times 12 =$ ___   77. $9 \times 0 =$ ___   78. $6 \times 6 =$ ___

79. $3 \times 7 =$ ___   80. $10 \times 2 =$ ___   81. $10 \times 10 =$ ___

82. $12 \times 7 =$ ___   83. $11 \times 4 =$ ___   84. $12 \times 1 =$ ___

85. $1 \times 7 =$ ___   86. $7 \times 6 =$ ___   87. $12 \times 4 =$ ___

88. $12 \times 5 =$ ___   89. $5 \times 4 =$ ___   90. $7 \times 0 =$ ___

MULTIPLES AND RULES

MULTIPLES

2s: 2, 4, 6, 8, 10, 12, 14, 16, 18, 20

4s: 4, 8, 12, 16, 20, 24, 28, 32, 36, 40

6s: 6, 12, 18, 24, 30, 36, 42, 48, 54, 60

8s: 8, 16, 24, 32, 40, 48, 56, 64, 72, 80

10s: 10, 20, 30, 40, 50, 60, 70, 80, 90, 100

12s: 12, 24, 36, 48, 60, 72, 84, 96, 108, 120

3s: 3, 6, 9, 12, 15, 18, 21, 24, 27, 30

5s: 5, 10, 15, 20, 25, 30, 35, 40, 45, 50

7s: 7, 14, 21, 28, 35, 42, 49, 56, 63, 70

9s: 9, 18, 27, 36, 45, 54, 63, 72, 81, 90

11s: 11, 22, 33, 44, 55, 66, 77, 88, 99, 110

MULTIPLICATION RULES

0s: Any number times 0 equals 0.

1s: Any number times 1 equals that same number.

2s: Any number times 2 is that number added to itself–the doubles in addition. All answers are even.

3s: Memorize answers by learning to count by three.

4s: All answers are even numbers. Answers have a repeating pattern.

    4   8 12 16 20
  24 28 32 36 40

5s: Five times an odd number ends in 5. Five times an even number ends in 0. When multiplying 5 times an even number, take half of the even number and add a zero.
Example: $5 \times 8 = ?$ Half of 8 is 4–the answer is 40.

6s: Use rules for other groups to remember 6s except for 6×7 and 6×8.

Recall	$6 \times 6 = 36$
Add another group of 6	$+\ 6$
	$6 \times 7 = 42$
Add another group of 6	$+\ 6$
	$6 \times 8 = 48$

7s: All are learned with other groups except 7×8. Learn this rule: 5, 6, 7, 8 or $56 = 7 \times 8$

8s: Use rules from other groups to remember 8s. Answers have repeating numerals in the 1s place.

    8 16 24 32 40
  48 56 64 72 80

9s: Nines answers are in pairs. 18 and 81, 27 and 72, 36 and 63, 45 and 54.

The number in the 10s place is one less than the number you are multiplying by 9. ($9 \times 3 = 27$, $9 \times 4 = 36$)

Numerals in one answer can be added together to equal 9. (18, $1 + 8 = 9$; 27, $2 + 7 = 9$; etc.)

Here is a rhyme:
When multiplying 9, keep this in mind: Look at a number in the answer line. Add the numerals together. They will equal nine.

10s: Add a zero to the right of the number. (Ex: $3 \times 10 = 30$)

11s: For a one-digit number, write that number in the 10s place and in the 1s place.

12s: For a one-digit number, write that number times 11 and then add that first number.

Ex:	$7 \times 12 =\ ?$
	$7 \times 11 = 77$
Add 7	$+\ 7$
	$7 \times 12 = 84$

TLC10112 Copyright © Teaching & Learning Company, Carthage, IL 62321-0010

FINDING EQUIVALENT FRACTIONS

Directions: Find the equivalent fractions.

1. $\frac{1 \times 3}{2 \times 3} = \frac{3}{6}$

2. $\frac{1 \times 2}{4 \times 2} =$ ___

3. $\frac{2 \times 3}{3 \times 3} =$ ___

4. $\frac{2 \times 5}{5 \times 5} =$ ___

5. $\frac{2 \times 4}{3 \times 4} =$ ___

6. $\frac{3 \times 6}{4 \times 6} =$ ___

7. $\frac{5 \times 3}{6 \times 3} =$ ___

8. $\frac{9 \times 3}{10 \times 3} =$ ___

Directions: Find the missing numerators.

9. $\frac{1}{4} = \frac{}{16}$

10. $\frac{3}{4} = \frac{}{16}$

11. $\frac{2}{3} = \frac{}{12}$

12. $\frac{4}{5} = \frac{}{10}$

13. $\frac{1}{7} = \frac{}{21}$

14. $\frac{2}{5} = \frac{}{25}$

15. $\frac{7}{8} = \frac{}{24}$

16. $\frac{9}{10} = \frac{}{50}$

Review

Watch the signs!

17. $\frac{1}{6}$
$+ \frac{3}{6}$

18. $\frac{3}{5}$
$+ \frac{1}{5}$

19. $\frac{10}{12}$
$- \frac{7}{12}$

20. $\frac{13}{15}$
$- \frac{6}{15}$

21. $\frac{21}{25}$
$- \frac{8}{25}$

22. $\frac{26}{30}$
$- \frac{19}{30}$

23. You have eight inches of ribbon left from an art project. What fractional part of a foot is this ribbon?

_____

24. You have five dimes and a nickel in your pocket. What fractional part of a dollar do you have?

_____

CHALLENGE:

In art class there are 12 girls and 13 boys. Six girls have blond hair. None of the boys have blond hair. What fractional part of the girls have blond hair? _____ What fractional part of the class has blond hair? _____ What fractional part of the boys have blond hair? _____

EQUIVALENT FRACTIONS

Directions: Write the multiples of these numbers.

3s: 3, 6, ___, ___, ___, ___, ___, ___, ___, ___

4s: 4, 8, ___, ___, ___, ___, ___, ___, ___, ___

6s: 6, 12, ___, ___, ___, ___, ___, ___, ___, ___

7s: 7, 14, ___, ___, ___, ___, ___, ___, ___, ___

8s: 8, 16, ___, ___, ___, ___, ___, ___, ___, ___

9s: 9, 18, ___, ___, ___, ___, ___, ___, ___, ___

Directions: Which fraction is larger? Use fraction strips. Circle the larger fraction.

1. $\frac{1}{3}$ or $\frac{1}{4}$ 2. $\frac{2}{5}$ or $\frac{3}{8}$ 3. $\frac{2}{6}$ or $\frac{3}{10}$

4. $\frac{3}{8}$ or $\frac{1}{3}$ 5. $\frac{3}{8}$ or $\frac{5}{12}$ 6. $\frac{4}{6}$ or $\frac{8}{10}$

Directions: Write in >, < or = signs. Use fraction strips to check.

7. $\frac{1}{3} \bigcirc \frac{1}{5}$ 8. $\frac{1}{4} \bigcirc \frac{1}{6}$ 9. $\frac{3}{8} \bigcirc \frac{2}{5}$

10. $\frac{5}{12} \bigcirc \frac{3}{8}$ 11. $\frac{9}{12} \bigcirc \frac{3}{4}$ 12. $\frac{2}{3} \bigcirc \frac{7}{10}$

Directions: Find the equivalent fractions.

13. $\frac{1}{4} = \frac{}{20}$ 14. $\frac{1}{3} = \frac{}{27}$ 15. $\frac{1}{5} = \frac{}{35}$ 16. $\frac{3}{10} = \frac{}{100}$

17. $\frac{3}{4} = \frac{}{24}$ 18. $\frac{4}{5} = \frac{}{30}$ 19. $\frac{7}{10} = \frac{}{80}$ 20. $\frac{3}{8} = \frac{}{56}$

21. You have a square fence to paint. If you have painted three sides, what fractional part have you painted? _____

22. You have $1/5$ of a watermelon–your friend has $2/12$. Who has more? _____

CHALLENGE:

You share your soft drink with a friend. You get $1/3$. She drinks $1/4$. What fractional part is left?
_____

(Note: Use fraction strips if you need help.)

Name _____

EQUIVALENT FRACTIONS

Directions: Write the multiples of these numbers.

3s: 3, 6, ____, ____, ____, ____, ____, ____, ____, ____

6s: 6, 12, ____, ____, ____, ____, ____, ____, ____

7s: 7, 14, ____, ____, ____, ____, ____, ____, ____

8s: 8, 16, ____, ____, ____, ____, ____, ____, ____

Directions: Find the equivalent fractions.

1. $\frac{5 \times 7}{6 \times 7}$ = ___

2. $\frac{7 \times 9}{8 \times 9}$ = ___

3. $\frac{5 \times 7}{7 \times 7}$ = ___

4. $\frac{8 \times 9}{9 \times 9}$ = ___

5. $\frac{6 \times 8}{7 \times 8}$ = ___

6. $\frac{9 \times 5}{11 \times 5}$ = ___

Directions: Find the missing numerators.

7. $\frac{3}{4} = \frac{}{24}$

8. $\frac{7}{8} = \frac{}{56}$

9. $\frac{9}{10} = \frac{}{70}$

10. $\frac{5}{6} = \frac{}{42}$

11. $\frac{6}{7} = \frac{}{63}$

12. $\frac{5}{9} = \frac{}{54}$

Review

Watch the signs.

13. $\frac{17}{27}$ $- \frac{9}{27}$

14. $\frac{21}{24}$ $- \frac{7}{24}$

15. $\frac{25}{30}$ $- \frac{18}{30}$

16. $\frac{37}{25}$ $- \frac{19}{25}$

17. $\frac{23}{50}$ $+ \frac{8}{50}$

18. $^4/_{15}$ of your class is sick. $^3/_{15}$ of your class is late. The rest are in their seats. What fraction of the students are in their seats? _____

19. You have three quarters, one dime, one nickel and two pennies. What fraction of a dollar do you have? _____

CHALLENGE:

You have 4/5 of a dollar. Your friend has 75¢. Who has more money? _____ How much more? _____

LESSON PLAN 4

REVIEW: Have the entire class chant each group of multiples.

INTRODUCTION: This step shows the opposite of the last step. Show students that just as they found equivalent fractions by multiplying both numerals in a fraction by the same number, they can divide both numerals by the same number.

DIVISION ASSESSMENT: Before assigning worksheets in this step, review division. Send home Parents' Note 4 (page 29) and the page of division facts (page 41) to study. If you do not teach Fraction Slapjack, cut off that part of the note.

Students will also need to practice finding common divisors. To help with this, teach students the Rules of Divisibility (page 40).

Students will not be able to work with fractions if they do not have a good knowledge of division facts and rules of divisibility.

REDUCING FRACTIONS: Use Equivalent Fraction Strips (page 28) to demonstrate reducing fractions.

Sometimes a fraction has a numerator and denominator that are large and hard to work with. We find a smaller fraction that is equivalent and that is called reducing. For example, have students look at the halves and the fourths fraction strips. What is another fraction that is equal to $2/4$? It is $1/2$.

We can change $2/4$ to $1/2$ by dividing the two and the four by 2.

$$\frac{2 \div 2}{4 \div 2} = \frac{1}{2}$$

In the fraction $5/10$, what number can divide evenly into both the 5 and 10? It is 5. Therefore,

$$\frac{5 \div 5}{10 \div 5} = \frac{1}{2}$$

Show with the fraction strip that $5/10 = 1/2$.

When a number will divide evenly into two numbers, it is called the common divisor. The common divisor for $5/10$ is 5. The common divisor for $2/4$ is 2.

What is the common divisor for $8/10$? Two. When you reduce, what do you get? $4/5$. Check with the fraction strips.

Try a few more examples if needed.

WALL CHART: Use Wall Chart C (page 42) to impress upon students that division by zero is not possible. Enlarge, laminate and post in the classroom or make a copy for each student.

FRACTION BINGO

1. Run off copies of the Fraction Bingo cards (page 46)–one card for each student. Have each student write *FREE* in a space of their choice.

2. Write these fractions on the board and have students write one fraction in each remaining space. They will not be able to use all fractions. Tell them to put fractions in random spaces so they will not all bingo at the same time.

   $1/2$, $1/3$, $1/4$, $1/5$, $1/6$, $1/7$, $1/8$, $4/7$, $3/8$, $5/7$, 0, $6/7$, $2/3$, $3/4$, $3/5$, $4/5$, $3/7$, $2/5$, $5/6$, $2/7$, 1, $5/8$, $7/8$, $1/9$, $4/9$, $5/9$, $7/9$, $8/9$, $3/10$, $7/10$, $9/10$

3. Laminate the cards. When the class plays bingo, students can X out fractions with a crayon and then clean off the Xs with a facial tissue when each game ends.

4. Cut apart the Fraction Bingo Calling Cards (pages 47-48). Call out the fraction and have the students X out the reduced fraction if it is on their cards. Have them use scratch paper if necessary to reduce fractions. (Example: You call $2/8$. Students X out $1/4$ on their cards.)

5. Ask parents to donate gum, candy, pencils, etc., for prizes.

FRACTION SLAPJACK

This is a two-player game. Teach this to students so they can play it at home or during recess. Only use class time to play games only once a month.

1. Use a standard deck of playing cards. Remove face cards, aces and sevens. We remove the face cards because they are confusing. Remove the aces and sevens because when they turn up, nothing will reduce them.

2. Deal out remaining cards into two piles facedown, one pile to each player.

3. Players turn the top card faceup at the same time. If the two numbers can be made into a fraction that can be reduced, call out the reduced fraction and slap the table before the other player can. (Use scratch paper if needed to see if the fraction was reduced correctly.) If someone calls out the wrong reduced fraction and slaps, the other player gets the cards. Use the smaller number as the numerator to form the fraction.

4. If the fraction cannot be reduced, turn over the next pair of cards. Continue until a pair of cards can be reduced. The winner of that pair gets all the cards faceup at that time. (Note: This same game can be played with the larger number as the numerator after students have learned about mixed numbers.)

5. If you can't locate free decks of cards to play this game, make copies of page 49–three copies for each student. These can be used like playing cards.

6. One person could play this game to practice reducing fractions alone. Also, two people could play it and take turns giving answers instead of competing for speed and slapping cards.

2s: Any number ending in 0, 2, 4, 6 or 8 is divisible by two.

3s: Add the numerals in a number. If they add up to 3, 6 or 9, the number is divisible by 3. (Example: 21: 2 + 1 = 3 and 21 is divisible by 3. 18: 1 + 8 = 9 and 18 is divisible by 3.)

4s: Every other even number is divisible by four. Memorize this pattern:

4	8	12	16	20
24	28	32	36	40

5s: Numbers ending in 0 or 5 are divisible by 5.

6s: Add the numerals in a number. If they add up to 3, 6 or 9 and the number is an even number, it is divisible by 6. (Example: 24: 2 + 4 = 6 and 24 is even–24 is divisible by 6.)

7s: Memorize the numbers divisible by seven.

8s: Numbers divisible by eight are always even. Memorize this pattern:

8	16	24	32	40
48	56	64	72	80

9s: Add the numerals in a number. If they add up to nine, the number is divisible by nine.

10s: Numbers ending in zero are divisible by 10.

Give the class some numbers and have them use these rules to find all the divisors of the numbers.

DIVISION FACTS

Zeros in Division–Zero Divided by Any Number Is Always Zero. This makes sense. If I have no money and I share with you, we both have no money. $(0 \div 2 = 0)$ If I share with nine people, we all have no money. $(0 \div 9 = 0)$

Division by Zero Is Undefined. Here's why:

In the problem:

$$\frac{5}{2 \overline{\smash{)}10}}$$

$2 \times 5 = 10$

In this problem:

$$\frac{?}{0 \overline{\smash{)}10}}$$

There is no number times zero that equals 10. For that reason we cannot divide by zero. Zero cannot be the denominator.

0 ÷ 1 = 0	0 ÷ 2 = 0	0 ÷ 3 = 0	0 ÷ 4 = 0	0 ÷ 5 = 0
1 ÷ 1 = 1	2 ÷ 2 = 1	3 ÷ 3 = 1	4 ÷ 4 = 1	5 ÷ 5 = 1
2 ÷ 1 = 2	4 ÷ 2 = 2	6 ÷ 3 = 2	8 ÷ 4 = 2	10 ÷ 5 = 2
3 ÷ 1 = 3	6 ÷ 2 = 3	9 ÷ 3 = 3	12 ÷ 4 = 3	15 ÷ 5 = 3
4 ÷ 1 = 4	8 ÷ 2 = 4	12 ÷ 3 = 4	16 ÷ 4 = 4	20 ÷ 5 = 4
5 ÷ 1 = 5	10 ÷ 2 = 5	15 ÷ 3 = 5	20 ÷ 4 = 5	25 ÷ 5 = 5
6 ÷ 1 = 6	12 ÷ 2 = 6	18 ÷ 3 = 6	24 ÷ 4 = 6	30 ÷ 5 = 6
7 ÷ 1 = 7	14 ÷ 2 = 7	21 ÷ 3 = 7	28 ÷ 4 = 7	35 ÷ 5 = 7
8 ÷ 1 = 8	16 ÷ 2 = 8	24 ÷ 3 = 8	32 ÷ 4 = 8	40 ÷ 5 = 8
9 ÷ 1 = 9	18 ÷ 2 = 9	27 ÷ 3 = 9	36 ÷ 4 = 9	45 ÷ 5 = 9
10 ÷ 1 = 10	20 ÷ 2 = 10	30 ÷ 3 = 10	40 ÷ 4 = 10	50 ÷ 5 = 10

0 ÷ 6 = 0	0 ÷ 7 = 0	0 ÷ 8 = 0	0 ÷ 9 = 0	0 ÷ 10 = 0
6 ÷ 6 = 1	7 ÷ 7 = 1	8 ÷ 8 = 1	9 ÷ 9 = 1	10 ÷ 10 = 1
12 ÷ 6 = 2	14 ÷ 7 = 2	16 ÷ 8 = 2	18 ÷ 9 = 2	20 ÷ 10 = 2
18 ÷ 6 = 3	21 ÷ 7 = 3	24 ÷ 8 = 3	27 ÷ 9 = 3	30 ÷ 10 = 3
24 ÷ 6 = 4	28 ÷ 7 = 4	32 ÷ 8 = 4	36 ÷ 9 = 4	40 ÷ 10 = 4
30 ÷ 6 = 5	35 ÷ 7 = 5	40 ÷ 8 = 5	45 ÷ 9 = 5	50 ÷ 10 = 5
36 ÷ 6 = 6	42 ÷ 7 = 6	48 ÷ 8 = 6	54 ÷ 9 = 6	60 ÷ 10 = 6
42 ÷ 6 = 7	49 ÷ 7 = 7	56 ÷ 8 = 7	63 ÷ 9 = 7	70 ÷ 10 = 7
48 ÷ 6 = 8	56 ÷ 7 = 8	64 ÷ 8 = 8	72 ÷ 9 = 8	80 ÷ 10 = 8
54 ÷ 6 = 9	63 ÷ 7 = 9	72 ÷ 8 = 9	81 ÷ 9 = 9	90 ÷ 10 = 9
60 ÷ 6 = 10	70 ÷ 7 = 10	80 ÷ 8 = 10	90 ÷ 9 = 10	100 ÷ 10 = 10

$0 \div 2 = 0$

$2\overline{)0}$ with 0 on top

$\dfrac{0}{2} = 0$

$0 \div 9 = 0$

$9\overline{)0}$ with 0 on top

$\dfrac{0}{9} = 0$

$2 \div 0$ (crossed out)

$0\overline{)2}$ (crossed out)

$\dfrac{2}{0}$ (crossed out)

$9 \div 0$ (crossed out)

$0\overline{)9}$ (crossed out)

$\dfrac{9}{0}$ (crossed out)

ZERO CANNOT BE THE DIVISOR.
ZERO CANNOT BE THE DENOMINATOR.

Name _____

DIVISION ASSESSMENT

1. $50 \div 5 =$ _____

2. $2 \div 2 =$ _____

3. $16 \div 8 =$ _____

4. $4 \div 4 =$ _____

5. $18 \div 9 =$ _____

6. $5 \overline{)25}$

7. $4 \overline{)12}$

8. $9 \div 9 =$ _____

9. $1 \div 1 =$ _____

10. $4 \div 2 =$ _____

11. $5 \overline{)20}$

12. $9 \overline{)27}$

13. $2 \overline{)6}$

14. $4 \overline{)36}$

15. $4 \overline{)0}$

16. $4 \overline{)16}$

17. $8 \overline{)48}$

18. $8 \overline{)8}$

19. $9 \overline{)36}$

20. $4 \overline{)8}$

21. $5 \overline{)15}$

22. $8 \overline{)24}$

23. $9 \overline{)45}$

24. $5 \overline{)0}$

25. $8 \overline{)32}$

26. $5 \overline{)45}$

27. $9 \overline{)54}$

28. $2 \overline{)8}$

29. $9 \overline{)63}$

30. $2 \overline{)10}$

31. $9 \overline{)72}$

32. $4 \overline{)28}$

33. $4 \overline{)20}$

34. $9 \overline{)81}$

35. $2 \overline{)16}$

36. $2 \overline{)12}$

37. $9 \overline{)0}$

38. $4 \overline{)32}$

39. $2 \overline{)14}$

40. 2) 18   41. 8) 56   42. 5) 40

43. 4) 24   44. 8) 0   45. 2) 0

46. 5) 30   47. 5) 35   48. 8) 64

49. 8) 80   50. 8) 72   51. 3) 6

52. 1) 0   53. 3) 12   54. 10) 70

55. 3) 3   56. 6) 6   57. 1) 9

58. 3) 9   59. 5) 5   60. 10) 20

61. 7) 56   62. 3) 15   63. 1) 2

64. 10) 10   65. 7) 7   66. 10) 0

67. 3) 18   68. 10) 80   69. 1) 3

70. 10) 30   71. 7) 49   72. 1) 8

73. 7) 63   74. 3) 21   75. 3) 27

76. 7) 42   77. 10) 40   78. 7) 0

44

DIVISION ASSESSMENT

79. $1\overline{)4}$ 80. $3\overline{)24}$ 81. $7\overline{)35}$

82. $10\overline{)90}$ 83. $1\overline{)6}$ 84. $10\overline{)60}$

85. $1\overline{)5}$ 86. $10\overline{)100}$ 87. $1\overline{)7}$

88. $6\overline{)0}$ 89. $6\overline{)36}$ 90. $3\overline{)0}$

91. $7\overline{)28}$ 92. $6\overline{)54}$ 93. $6\overline{)42}$

94. $6\overline{)12}$ 95. $6\overline{)24}$ 96. $7\overline{)21}$

97. $6\overline{)48}$ 98. $6\overline{)18}$ 99. $7\overline{)14}$

100. $6\overline{)60}$ 101. $4\overline{)40}$ 102. $5\overline{)10}$

103. $9\overline{)90}$ 104. $8\overline{)40}$ 105. $2\overline{)20}$

106. $3\overline{)30}$ 107. $1\overline{)10}$ 108. $6\overline{)30}$

109. $7\overline{)70}$ 110. $10\overline{)50}$

FRACTION BINGO

$\dfrac{3}{6}$	$\dfrac{2}{6}$	$\dfrac{2}{8}$	$\dfrac{3}{15}$	$\dfrac{2}{12}$
$\dfrac{2}{14}$	$\dfrac{2}{16}$	$\dfrac{8}{14}$	$\dfrac{6}{16}$	$\dfrac{15}{21}$
$\dfrac{0}{5}$	$\dfrac{12}{14}$	$\dfrac{4}{6}$	$\dfrac{6}{8}$	$\dfrac{6}{10}$
$\dfrac{8}{10}$	$\dfrac{6}{14}$	$\dfrac{4}{10}$	$\dfrac{10}{12}$	$\dfrac{4}{14}$

$\dfrac{10}{10}$	$\dfrac{10}{16}$	$\dfrac{21}{24}$	$\dfrac{2}{18}$	$\dfrac{8}{18}$
$\dfrac{10}{18}$	$\dfrac{14}{18}$	$\dfrac{16}{18}$	$\dfrac{6}{20}$	$\dfrac{14}{20}$
$\dfrac{18}{20}$				

1	2	3	4	5
6	7	8	9	10
1	2	3	4	5
6	7	8	9	10

FRACTION SLAPJACK CARDS

REDUCING FRACTIONS

Directions: Reduce these fractions.

1. $\frac{5 \div 5}{10 \div 5} = \frac{1}{2}$

2. $\frac{4 \div 2}{6 \div 2} = $ ___

3. $\frac{3 \div 3}{12 \div 3} = $ ___

4. $\frac{16 \div 4}{20 \div 4} = $ ___

5. $\frac{4 \div 2}{14 \div 2} = $ ___

6. $\frac{6 \div 6}{12 \div 6} = $ ___

Directions: Reduce. Find a common divisor.

7. $\frac{6}{10} = \frac{}{5}$

8. $\frac{4}{12} = \frac{}{3}$

9. $\frac{8}{18} = \frac{}{9}$

10. $\frac{20}{30} = \frac{}{3}$

Directions: Reduce.

11. $\frac{7}{28} = $ ___

12. $\frac{14}{16} = $ ___

13. $\frac{15}{25} = $ ___

14. $\frac{18}{27} = $ ___

15. $\frac{24}{36} = $ ___

Directions: Add or subtract. Reduce the answer if you can.

16. $\frac{1}{12}$
$+ \frac{3}{12}$

17. $\frac{1}{6}$
$+ \frac{3}{6}$

18. $\frac{9}{10}$
$- \frac{4}{10}$

19. $\frac{3}{4}$
$- \frac{1}{4}$

20. $\frac{14}{15}$
$- \frac{8}{15}$

21. If you eat $^1/_6$ of your pizza today and $^2/_6$ of your pizza tomorrow, how much will you have eaten in all?

_____

22. If $^2/_5$ of the class got a B on the spelling test and $^3/_5$ of the class got a C on the test, what is the total fraction of the class that got a B or C?

_____

CHALLENGE:

Two friends want to buy your newest baseball trading card. One will pay $^{12}/_4$ dollars and the other will pay $^{12}/_3$ dollars. Which is the better price? _____

TREASURE HUNT

Reduce all the fractions. Then find two paths to the treasure through fractions that are equal.

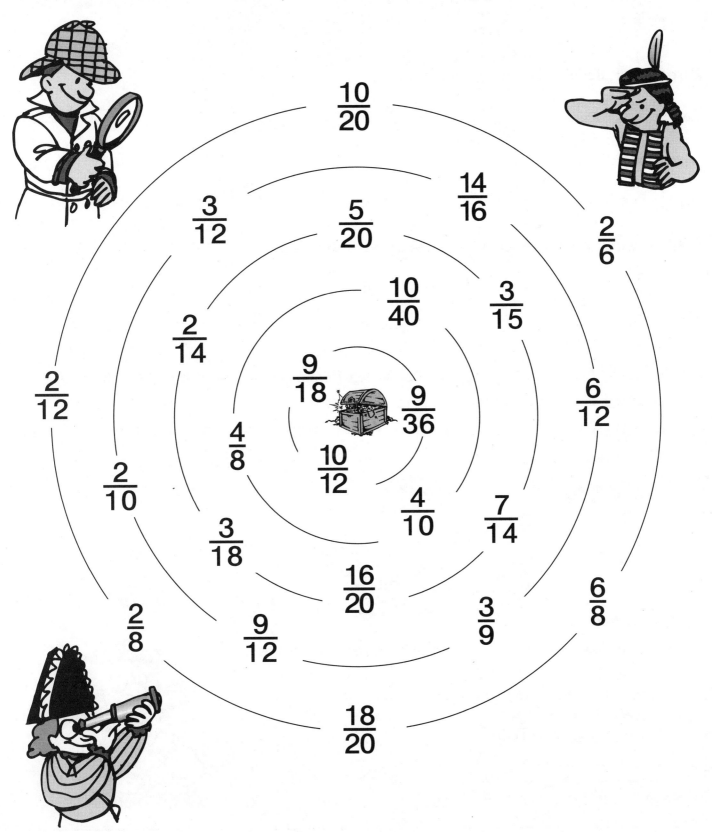

REDUCING FRACTIONS

Directions: Reduce these fractions.

1. $\frac{15}{30}$ = ____   2. $\frac{25}{40}$ = ____   3. $\frac{18}{72}$ = ____   4. $\frac{56}{64}$ = ____   5. $\frac{90}{100}$ = ____

Directions: Reduce. Find a common divisor.

6. $\frac{24}{36}$ = ____   7. $\frac{54}{60}$ = ____   8. $\frac{72}{81}$ = ____   9. $\frac{70}{90}$ = ____   10. $\frac{32}{40}$ = ____

Directions: Reduce.

11. $\frac{21}{35}$ = ____   12. $\frac{18}{30}$ = ____   13. $\frac{30}{80}$ = ____   14. $\frac{28}{63}$ = ____   15. $\frac{45}{50}$ = ____

Directions: Add or subtract. Reduce the answer if you can.

16. $\frac{9}{15}$ 17. $\frac{6}{14}$ 18. $\frac{18}{33}$ 19. $\frac{17}{50}$ 20. $\frac{23}{50}$

$+\ \frac{3}{15}$ $-\ \frac{2}{14}$ $+\ \frac{14}{33}$ $+\ \frac{13}{50}$ $+\ \frac{19}{50}$

21. You have $^9/_{10}$ of a yard of wallpaper border left from papering your room. You will use $^1/_2$ of a yard for a craft project. Then how much will be left? _____

22. You have spent 33 minutes eating dinner. What fractional part of an hour is that? _____

CHALLENGE:

Here is a message for you. Fill in the blanks to read the message.

a. first half of the word GOAL.
b. second half of the word WOOD.
c. first $^2/_3$ of the word JOG.
d. last $^1/_3$ of the word TUB.

____ ____   ____ ____!
  a   b   c   d

REVIEW: Continue to have the class chant the multiples until students have them memorized. You could have students pair off and say the multiples to each other or have a parent volunteer listen to them. Knowing these multiples is very important.

INTRODUCTION: Adding and subtracting fractions is not always as simple as $1/3 + 1/3 = 2/3$. Some fractions do not have the same denominator. We need the same denominator in order to add two fractions. The denominator is like a label. There is an old saying that you can't add apples and oranges. If you have 5 apples and 6 oranges and you add them, you could say you have 11 apple-oranges, but that is confusing. If you change the label to *fruit*, then it makes more sense to say you have 11 pieces of fruit. It's the same with fractions. We need to write fractions with the same label or denominator. Then we can add them and know what we have.

Example: You have two jugs of milk which are partly full. One has $1/4$ of a gallon and the other has $1/2$ of a gallon. How much of a gallon will you have when they are combined?

We know from learning equivalent fractions that $1/2 = 2/4$. Now we can add

$$\frac{1}{2} = \frac{2}{4}$$
$$+\frac{1}{4} = \frac{1}{4}$$
$$\frac{3}{4}$$

Changing a fraction to an equivalent fraction so you can add or subtract is called **finding a common denominator**.

COMMON DENOMINATORS: Here are two simple rules for finding common denominators.

1. See if the larger denominator is a multiple of the smaller one. If it is, that is the common denominator. The $1/4$ and $1/2$ examples show this.

2. Multiply the two denominators to get a common denominator.

   For $1/4 + 1/5$ you cannot use the larger denominator because 5 is not a multiple of 4. $4 \times 5 = 20$ and 20 will work.

   $$\frac{1}{4} = \frac{5}{20} \qquad \frac{1}{5} = \frac{4}{20}$$

Now you can add

$$\frac{1}{4} = \frac{5}{20}$$
$$+\frac{1}{5} = \frac{4}{20}$$
$$\frac{9}{20}$$

Students will practice with these two rules for finding common denominators now and learn about larger numbers later.

MULTIPLES: You can use the page of multiples (page 34) to show students how the rules work. If you are looking for a common denominator for 4 and 5, look at the list of multiples and find the smallest number that is in both lists. Give them a few number pairs for practice. (3 and 6, 7 and 6, 6 and 8, etc.)

Students should learn the multiples and not use the page of multiples as a crutch.

OPTIONAL: You could enlarge and laminate the Wall Chart D (page 56) with the common denominator rules.

OPTIONAL: For the level 2 worksheet (page 59), you need to show students how to borrow from the whole number for problems 11-15.

Example:

$$5\frac{1}{10}$$
$$-\ 2\frac{7}{10}$$

You can't subtract $^7/_{10}$ from $^1/_{10}$. You can borrow from the whole number 5. Since you are using tenths, borrow $^{10}/_{10}$ from the 5.

$$\overset{4}{\cancel{5}}\frac{1}{10} + \frac{10}{10} = \frac{11}{10}$$
$$-\ 2\frac{7}{10} \qquad \frac{7}{10}$$
$$\overline{2} \qquad \overline{\frac{4}{10} = 2\frac{4}{10} = 2\frac{2}{5}}$$

Now you can subtract. Show them one more example.

$$\overset{2}{\cancel{3}}\frac{1}{3} + \frac{3}{3} = \frac{4}{3}$$
$$-\ 1\frac{2}{3} \qquad \frac{2}{3}$$
$$\overline{1} \qquad \overline{\frac{2}{3} = 1\frac{2}{3}}$$

Also, show how to find common denominators for adding three fractions.

$$\frac{1}{2} = \frac{}{10}$$
$$\frac{1}{5} = \frac{}{10}$$
$$+\ \frac{3}{10} = \frac{}{10}$$

FRACTION ACTION

FRACTION ACTION: 2-6 players

Take five to 10 minutes to teach the class Fraction Action. This game is harder to play than Fraction Bingo or Fraction Slapjack. You might only have a few in the class who will want to play this, but the game is good practice for finding common denominators.

Teach this to the entire class although it might be too difficult for some of your students. It could make a good challenge for some of the kids who always get their homework done quickly. Use standard playing cards. Ask parents to donate old decks of cards. If you can't locate decks of cards, copy page 49, on card stock if possible, for this game.

1. Deal one card facedown and one card faceup to each player. Players may then ask for another card faceup or pass and get no more cards. Players may have up to eight cards if they wish.

2. For each card dealt, think of the fraction using the number on the card as the denominator. (Example: If you have a 2, it is worth $1/2$. For a 3, the value is $1/3$. Aces and face cards are each worth $1/10$.)

3. The object of the game is to get the total of the fractions closest to one without going over. (Example: If you get two 2s, $1/2 + 1/2 = 1$ and you win. If you get three 3s, $1/3 + 1/3 + 1/3 = 1$ and you win. If you get a 2 and a 4, $1/2 + 1/4 = 3/4$ and if no one gets more than $3/4$, you win. If you get a 2, a 2 and a 4, $1/2 + 1/2 + 1/4 = 1 1/4$, you lose.)

4. The winner for each hand gets 1 point. It is possible to have more than one winner for each hand. First person to get 10 points wins. (Use scratch paper to check answers.)

$$\frac{1}{4} + \frac{1}{2} + \frac{1}{8}$$

$$\frac{1}{3} + \frac{1}{7} +$$

RULES FOR FINDING A COMMON DENOMINATOR

If the larger denominator is a multiple of the smaller one, use the larger as the common denominator.

Example:

$$\frac{1}{2} = \frac{2}{4}$$
$$+\frac{1}{4} = \frac{1}{4}$$
$$\overline{\frac{3}{4}}$$

If the larger denominator is *not* a multiple of the smaller one, multiply the two denominators (3 x 5 = 15) to find the common denominator.

Example:

$$\frac{1}{3} = \frac{5}{15}$$
$$+\frac{1}{5} = \frac{3}{15}$$
$$\overline{\frac{8}{15}}$$

56

COMMON DENOMINATORS

Directions: Add. Find the common denominators. Use the larger denominator as the common denominator. Reduce if you can.

1. $\frac{1}{2}$ $\begin{smallmatrix} \times 2 \\ \times 2 \end{smallmatrix}$ = $\frac{}{4}$

   + $\frac{1}{4}$ = $\frac{}{4}$

2. $\frac{3}{4}$ = ___

   + $\frac{7}{8}$ = ___

3. $\frac{1}{2}$ = ___

   + $\frac{1}{6}$ = ___

4. $\frac{2}{5}$ = ___

   + $\frac{1}{10}$ = ___

5. $\frac{2}{3}$ = ___

   + $\frac{1}{6}$ = ___

6. $\frac{4}{5}$ = ___

   + $\frac{1}{10}$ = ___

7. $\frac{1}{8}$ = ___

   + $\frac{3}{16}$ = ___

8. $\frac{3}{4}$ = ___

   + $\frac{1}{12}$ = ___

Directions: Subtract. Use the larger denominator as the common denominator. Reduce if you can.

9. $\frac{1}{2}$ = ___

   - $\frac{1}{8}$ = ___

10. $\frac{3}{4}$ = ___

    - $\frac{3}{8}$ = ___

11. $\frac{2}{3}$ = ___

    - $\frac{1}{6}$ = ___

12. $\frac{3}{5}$ = ___

    - $\frac{1}{10}$ = ___

13. $\frac{2}{3}$ = ___

    - $\frac{1}{12}$ = ___

14. $\frac{5}{6}$ = ___

    - $\frac{1}{3}$ = ___

15. $\frac{5}{16}$ = ___

    - $\frac{1}{8}$ = ___

16. $\frac{3}{5}$ = ___

    - $\frac{2}{25}$ = ___

Directions: Add or subtract. Multiply denominators to find the common denominator. Reduce if you can.

17. $\frac{1}{3}$ = ___

    + $\frac{1}{4}$ = ___

18. $\frac{1}{3}$ = ___

    - $\frac{1}{4}$ = ___

19. $\frac{2}{5}$ = ___

    - $\frac{1}{4}$ = ___

20. $\frac{2}{3}$ = ___

    - $\frac{1}{5}$ = ___

21. $\frac{2}{3}$ = ___

    + $\frac{1}{2}$ = ___

22. $\frac{4}{5}$ = ___

    - $\frac{1}{4}$ = ___

23. $\frac{1}{5}$ = ___

    + $\frac{1}{6}$ = ___

24. $\frac{1}{6}$ = ___

    + $\frac{1}{7}$ = ___

CHALLENGE:

Your snack mix calls for $\frac{1}{2}$ cup raisins, $\frac{2}{3}$ cup nuts and $\frac{1}{4}$ cup chocolate chips. What is the total of these three items? _____

Name _____

COMMON DENOMINATORS

RULE 1: Use the larger denominator for the common denominator.

RULE 2: Multiply the two denominators for the common denominator.

Directions: Add. Find the common denominators. Use the rules that work best. Reduce if you can.

1. $\frac{7}{10}$ = ___   2. $\frac{3}{9}$ = ___   3. $\frac{3}{4}$ = ___   4. $\frac{2}{3}$ = ___

  $+ \frac{1}{2}$ = ___ $+ \frac{1}{2}$ = ___ $+ \frac{1}{8}$ = ___ $+ \frac{1}{6}$ = ___

5. $\frac{3}{7}$ = ___   6. $\frac{5}{8}$ = ___   7. $\frac{2}{5}$ = ___   8. $\frac{3}{11}$ = ___

  $+ \frac{3}{14}$ = ___ $+ \frac{3}{16}$ = ___ $+ \frac{3}{6}$ = ___ $+ \frac{1}{2}$ = ___

Directions: Subtract. Find the common denominators. Use the rules that work best. Reduce if you can.

9. $\frac{1}{2}$ = ___   10. $\frac{6}{7}$ = ___   11. $\frac{2}{3}$ = ___   12. $\frac{2}{3}$ = ___

  $- \frac{1}{4}$ = ___ $- \frac{1}{3}$ = ___ $- \frac{1}{4}$ = ___ $- \frac{3}{5}$ = ___

13. $\frac{1}{6}$ = ___   14. $\frac{4}{9}$ = ___   15. $\frac{1}{2}$ = ___   16. $\frac{7}{8}$ = ___

  $- \frac{1}{18}$ = ___ $- \frac{1}{4}$ = ___ $- \frac{4}{9}$ = ___ $- \frac{3}{9}$ = ___

Directions: Add or subtract. Find the common denominators. Reduce if you can.

17. $\frac{3}{7}$ = ___   18. $\frac{5}{6}$ = ___   19. $\frac{3}{4}$ = ___   20. $\frac{3}{4}$ = ___

  $+ \frac{1}{3}$ = ___ $- \frac{5}{12}$ = ___ $- \frac{4}{9}$ = ___ $+ \frac{1}{8}$ = ___

21. $\frac{3}{10}$ = ___   22. $\frac{4}{5}$ = ___   23. $\frac{2}{5}$ = ___   24. $\frac{3}{7}$ = ___

  $+ \frac{1}{3}$ = ___ $- \frac{1}{2}$ = ___ $+ \frac{1}{4}$ = ___ $+ \frac{1}{2}$ = ___

CHALLENGE:

Gym class jogged $2\frac{1}{2}$ miles and then walked $1\frac{3}{4}$ miles. What is the total distance? _____

Name _____

COMMON DENOMINATORS

RULE 1: Use the larger denominator for the common denominator.

RULE 2: Multiply the two denominators for the common denominator.

Directions: Add. Use the rule that works best. Reduce if you can.

1. $\frac{1}{2}$ = ___   2. $\frac{1}{3}$ = ___   3. $\frac{1}{3}$ = ___   4. $\frac{1}{2}$ = ___   5. $\frac{1}{3}$ = ___

    $\frac{1}{4}$ = ___ $\frac{1}{6}$ = ___ $\frac{1}{2}$ = ___ $\frac{1}{5}$ = ___ $\frac{1}{5}$ = ___

+ $\frac{1}{8}$ = ___   + $\frac{1}{12}$ = ___   + $\frac{1}{6}$ = ___   + $\frac{1}{10}$ = ___   + $\frac{1}{15}$ = ___

6. $\frac{2}{5}$ = ___   7. $\frac{3}{4}$ = ___   8. $\frac{2}{3}$ = ___   9. $\frac{2}{3}$ = ___   10. $\frac{1}{3}$ = ___

    $\frac{1}{2}$ = ___ $\frac{1}{2}$ = ___ $\frac{1}{2}$ = ___ $\frac{5}{6}$ = ___ $\frac{4}{5}$ = ___

+ $\frac{7}{10}$ = ___   + $\frac{5}{8}$ = ___   + $\frac{5}{6}$ = ___   + $\frac{11}{12}$ = ___   + $\frac{11}{15}$ = ___

Directions: Subtract. Borrow from the whole number. Reduce if you can.

11. $11\frac{7}{10}$ = ___   12. $12\frac{5}{6}$ = ___   13. $5\frac{1}{3}$ = ___   14. $10\frac{1}{10}$ = ___   15. $15\frac{3}{7}$ = ___

   - $7\frac{4}{5}$ = ___   - $3\frac{11}{12}$ = ___   - $2\frac{2}{3}$ = ___   - $3\frac{7}{10}$ = ___   - $8\frac{6}{7}$ = ___

16. Thirty students planned to go the library, but only 24 went. What fraction actually went? Reduce.

_____

17. You will need $5\frac{1}{2}$ yards of fabric for a tent. You have $4\frac{7}{8}$ yards. How much more will you need?

_____

18. If you are $52\frac{1}{2}$ inches tall and your best friend is $48\frac{3}{4}$ inches tall, how much taller are you?

_____

CHALLENGE:

Write each letter in the space above its equivalent fraction to answer this question: Which day of the year is a complete sentence?

$A = \frac{3}{9}$; $C = \frac{3}{24}$; $F = \frac{4}{16}$; $H = \frac{2}{12}$; $M = \frac{2}{20}$;

$O = \frac{2}{14}$; $R = \frac{3}{15}$; $T = \frac{6}{12}$; $U = \frac{2}{24}$

____ ____ ____ ____ ____

$\frac{1}{10}$ $\frac{1}{3}$ $\frac{1}{5}$ $\frac{1}{8}$ $\frac{1}{6}$

____ ____ ____ ____ ____ ____

$\frac{1}{4}$ $\frac{1}{7}$ $\frac{1}{12}$ $\frac{1}{5}$ $\frac{1}{2}$ $\frac{1}{6}$

TLC10112 Copyright © Teaching & Learning Company, Carthage, IL 62321-0010

LEVEL 2

59

LESSON PLAN 6

INTRODUCTION: So far we have had only fractions less than one—except for a couple of Challenge problems. These are fractions with the numerator smaller than the denominator, and they are called proper fractions. Of course, in real-life situations, people have to work with larger fractions. Here students will begin to work problems with numbers larger than one, numbers in which the numerator is larger than the denominator.

Proper fractions are fractions which have a numerator smaller than the denominator. An example is $3/4$.

Improper fractions are fractions which have a numerator larger than the denominator. An example is $5/3$.

IMPROPER FRACTIONS: To demonstrate improper fractions, you could draw circles on the board or use the Fraction Circles included on pages **66-68**. Consider the fraction $5/3$. Draw two circles on the board and divide each one into thirds. Shade in five sections. Ask the students to think of another way to write $5/3$. They will see that they have one complete circle with $2/3$ left over.

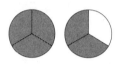

Thus $5/3 = 1\frac{2}{3}$.

Try another example—$10/3$. Ask if they can figure out how many circles this would give and how many thirds would be left over. Since each circle would give $3/3$, three circles would give $9/3$ and there would be $1/3$ left over. We don't want to be drawing circles all the time, and what we really want to know is how many groups of 3 are in 10, so we divide. In Lesson Plan 1, we said that a fraction means to divide and here is a time when division saves time. $10/3 = 3\frac{1}{3}$.

$$3\overline{)\begin{array}{r} 3 \\ 10 \\ 9 \\ \hline 1 \end{array}} \quad = \quad 3\frac{1}{3}$$

Go ahead and draw four circles so that students will be convinced that this works.

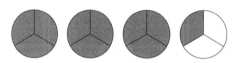

Try a few more examples. Draw the circles and then get the same answer by dividing. Try $7/2$, $11/3$ and $11/4$. When the answer contains both a whole number and a fraction, it is called a mixed number.

A mixed number is a number containing a whole number and a fraction.

We talk about changing from improper fractions to mixed numbers. We can also change from a mixed number to an improper fraction. We will work with that in the next section. (Show level 2 students how to do this before assigning the level 2 worksheet.)

60

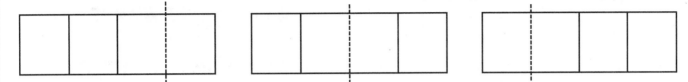

BONUS: In Lesson Plan 1 (pages 9-10) and Wall Chart A (page 13) we learned that the fraction ³/₄ means something has been divided into 4 equal parts and we are talking about 3 of them. ³/₄ also means 3 objects divided into four equal parts. If you had only 3 candy bars and 4 kids to share them, here is what you would find:

When they share the pieces equally, each kid would get ³/₄ of a candy bar. This might be confusing to some of your students, but some students love playing with ideas like this.

REDUCING SHORTCUT: Students have already had practice with reducing so this is a good time to show them a time-saving trick for reducing numbers divisible by 10 or 100.

Show them this fraction: ¹⁰/₅₀

We know that 10 and 50 are both divisible by 10 and ¹⁰/₅₀ = ¹/₅.

What about ¹⁰⁰/₅₀₀? 100 and 500 are both divisible by 100. When we reduce, we get ¹/₅. Instead of going through the process of dividing by 10 or 100, we can just think of crossing off a zero in both the numerator and denominator . . . or think of crossing off two zeros in both the numerator and denominator.

Try a few more for practice. ⁷⁰/₉₀, ²⁰/₅₀, ³⁰⁰/₇₀₀, ²⁰⁰/₃₀₀.

IMPROPER FRACTION SLAPJACK: Students can use the same rules for Fraction Slapjack (page 39) except to use the larger number as the numerator and then change the improper fraction to a mixed number. For example, if the two cards turned over were 6 and 4, make the fraction ⁶/₄ and reduce to 1²/₄ = 1¹/₂. All other rules stay the same.

FRACTIONS

PROPER FRACTION

$$\frac{3}{4} = \frac{\text{SMALLER NUMBER}}{\text{LARGER NUMBER}}$$

IMPROPER FRACTION

$$\frac{7}{2} = \frac{\text{LARGER NUMBER}}{\text{SMALLER NUMBER}}$$

MIXED NUMBER

$$3\frac{1}{2} = \frac{\text{WHOLE NUMBER}}{\text{AND A FRACTION}}$$

MULTIPLICATION OF FRACTIONS

$$\frac{1}{4} \times \frac{1}{3} = \frac{1}{12}$$

(MULTIPLY NUMERATORS)

(MULTIPLY DENOMINATORS)

DIVISION OF FRACTIONS

$$\frac{2}{3} \div \frac{1}{3}$$ INVERT THE SECOND FRACTION AND MULTIPLY

$$\frac{2}{3} \times \frac{3}{1} = \frac{6}{3} = 2$$

IMPROPER FRACTIONS

Directions: Change these improper fractions to mixed numbers by dividing.

1. $\frac{4}{3}$ = ___   2. $\frac{6}{5}$ = ___   3. $\frac{10}{7}$ = ___   4. $\frac{11}{8}$ = ___

5. $\frac{20}{9}$ = ___   6. $\frac{17}{6}$ = ___   7. $\frac{15}{4}$ = ___   8. $\frac{19}{9}$ = ___

Review

Directions: Reduce.

9. $\frac{7}{14}$ = ___   10. $\frac{12}{20}$ = ___   11. $\frac{10}{25}$ = ___   12. $\frac{14}{21}$ = ___

13. $\frac{18}{45}$ = ___   14. $\frac{30}{60}$ = ___   15. $\frac{28}{63}$ = ___   16. $\frac{9}{24}$ = ___

Directions: Find the equivalent fractions.

17. $\frac{1}{3} = \frac{}{15}$ 18. $\frac{3}{4} = \frac{}{16}$ 19. $\frac{9}{10} = \frac{}{50}$ 20. $\frac{3}{7} = \frac{}{35}$

21. $\frac{3}{8} = \frac{}{72}$ 22. $\frac{2}{10} = \frac{}{90}$ 23. $\frac{2}{9} = \frac{}{54}$ 24. $\frac{5}{7} = \frac{}{56}$

Directions: Add or subtract. Find the common denominators. Reduce if you can.

25. $\frac{1}{4}$ = ___   26. $\frac{1}{3}$ = ___   27. $\frac{3}{5}$ = ___   28. $\frac{3}{4}$ = ___

$+ \frac{1}{12}$ = ___ $+ \frac{5}{9}$ = ___ $- \frac{1}{6}$ = ___ $- \frac{3}{12}$ = ___

29. $\frac{4}{9}$ = ___   30. $\frac{2}{5}$ = ___   31. $\frac{11}{12}$ = ___   32. $\frac{3}{7}$ = ___

$- \frac{1}{3}$ = ___ $- \frac{3}{15}$ = ___ $- \frac{1}{4}$ = ___ $- \frac{2}{5}$ = ___

$\frac{1}{3}$ $\frac{1}{4}$ $\frac{2}{5}$ $\frac{3}{7}$

CHALLENGE:

You need 9/5 feet of wood trim for the art class project. It costs $3 per foot. What is the total cost? _____

IMPROPER FRACTIONS

Directions: Matching. Put the letter of the correct fraction in the space before each word.

1. _____ Mixed Number   A. $\frac{⑨}{10}$

2. _____ Numerator (circled)   B. $3\frac{7}{8}$

3. _____ Proper Fraction   C. $\frac{3}{4}$

4. _____ Denominator (circled)   D. $\frac{12}{3}$

5. _____ Improper Fraction   E. $\frac{1}{④}$

Directions: Reduce. Remember the shortcut for 10s and 100s.

6. $\frac{10}{20}$ = ___   7. $\frac{30}{60}$ = ___   8. $\frac{20}{50}$ = ___   9. $\frac{80}{90}$ = ___

10. $\frac{100}{200}$ = ___   11. $\frac{300}{600}$ = ___   12. $\frac{200}{500}$ = ___   13. $\frac{10}{100}$ = ___

Directions: Change to mixed numbers. Reduce if possible.

14. $\frac{6}{3}$ = ___   15. $\frac{13}{6}$ = ___   16. $\frac{24}{9}$ = ___   17. $\frac{10}{4}$ = ___

18. $\frac{16}{7}$ = ___   19. $\frac{14}{5}$ = ___   20. $\frac{10}{2}$ = ___   21. $\frac{14}{8}$ = ___

Directions: Add. Reduce if possible. Some must be changed to equivalent fractions.

22. $\frac{5}{10}$ 23. $\frac{3}{8}$ 24. $\frac{5}{6}$ 25. $\frac{1}{4}$ 26. $\frac{7}{8}$
$+\frac{3}{10}$ $+\frac{5}{8}$ $+\frac{5}{6}$ $+\frac{5}{6}$ $+\frac{3}{4}$

27. $\frac{9}{10}$ 28. $\frac{6}{7}$ 29. $\frac{9}{10}$ 30. $\frac{5}{8}$
$+\frac{3}{4}$ $+\frac{7}{8}$ $+\frac{3}{6}$ $+\frac{4}{9}$

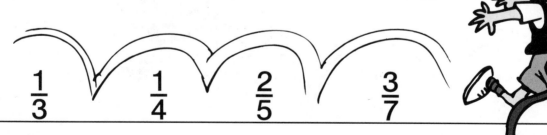

$\frac{1}{3}$ $\frac{1}{4}$ $\frac{2}{5}$ $\frac{3}{7}$

CHALLENGE:

You worked 1½ hours on yard work and 1⅔ hours on housework. Your mom paid you $3 per hour. How much did you earn? _____

Name _____

IMPROPER FRACTIONS

Directions: Change to mixed numbers. Show your work on the back.

1. $\frac{53}{7}$ = ___

2. $\frac{47}{13}$ = ___

3. $\frac{33}{17}$ = ___

4. $\frac{61}{18}$ = ___

5. $\frac{37}{18}$ = ___

6. $\frac{29}{13}$ = ___

7. $\frac{23}{18}$ = ___

8. $\frac{37}{15}$ = ___

9. $\frac{49}{16}$ = ___

10. $\frac{41}{19}$ = ___

11. $\frac{57}{16}$ = ___

12. $\frac{59}{23}$ = ___

Directions: Reduce. Use rules of divisibility.

13. $\frac{13}{39}$ = ___

14. $\frac{17}{51}$ = ___

15. $\frac{11}{33}$ = ___

16. $\frac{20}{460}$ = ___

17. $\frac{200}{600}$ = ___

18. $\frac{20}{1000}$ = ___

19. $\frac{27}{81}$ = ___

20. $\frac{54}{96}$ = ___

Directions: Change to improper fractions.

21. $19\frac{1}{7}$ = ___

22. $35\frac{1}{5}$ = ___

23. $48\frac{1}{10}$ = ___

24. $37\frac{1}{4}$ = ___

25. If normal body temperature is $98^{6}/_{10}$ degrees and your temperature is $2^{1}/_{2}$ degrees above normal, what is your temperature? _____

26. Your teacher says there are $^{96}/_{8}$ school days left. How many days are left? _____

CHALLENGE:

Change to improper fractions to answer this question: When is the best time to visit the dentist?

$$H = \frac{21}{10}; \quad I = \frac{13}{5}; \quad O = \frac{7}{2}; \quad R = \frac{9}{2}; \quad T = \frac{5}{3}; \quad W = \frac{17}{4}; \quad Y = \frac{17}{6}$$

___ ___ ___   ___   ___ ___ ___ ___ ___ ___
$1\frac{2}{3}$ $4\frac{1}{4}$ $3\frac{1}{2}$ $1\frac{2}{3}$ $2\frac{1}{10}$ $2\frac{3}{5}$ $4\frac{1}{2}$ $1\frac{2}{3}$ $2\frac{5}{6}$

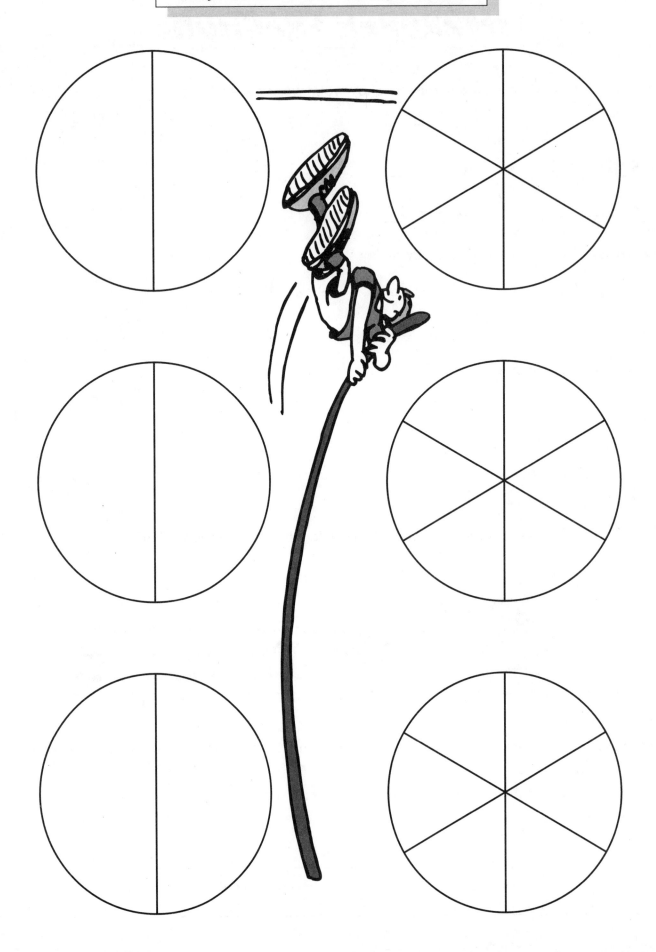

FRACTION CIRCLES

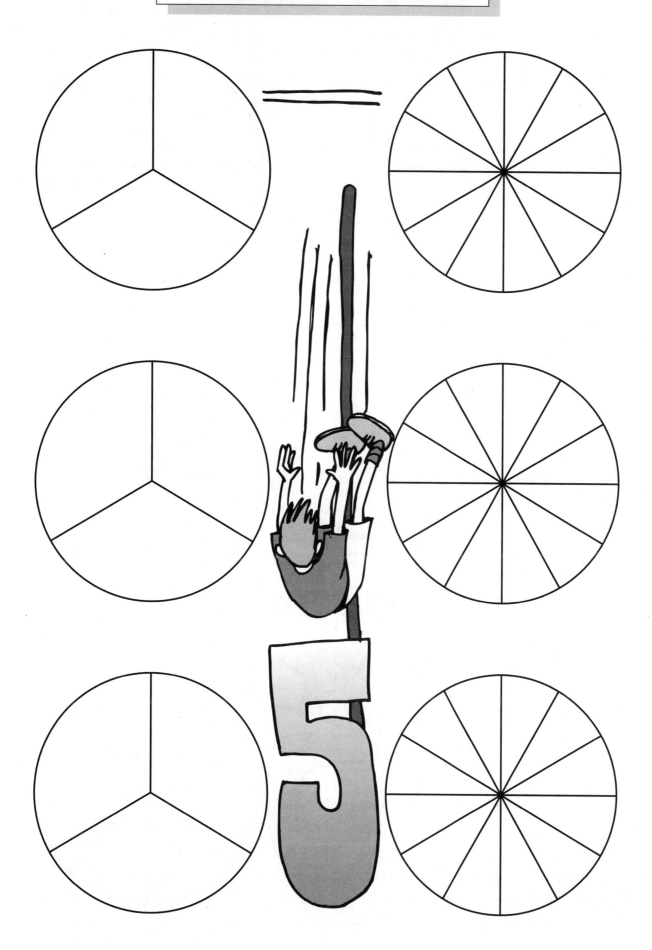

FRACTION CIRCLES

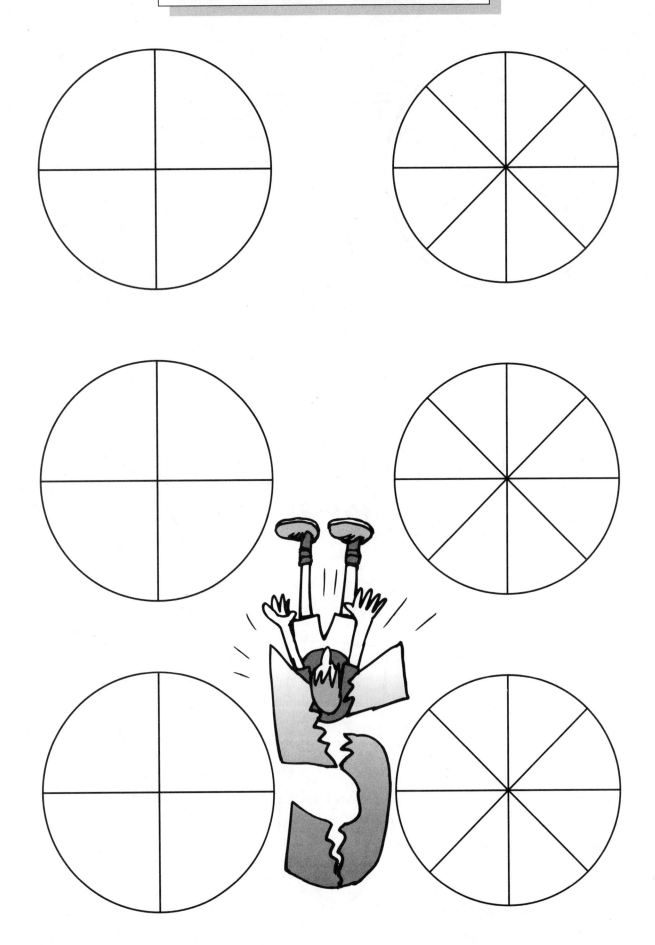

LESSON PLAN 7

INTRODUCTION: Here students will learn the reverse of lesson 6. We will begin with a mixed number like $1\frac{1}{3}$ and show students how to change it to an improper fraction—$\frac{4}{3}$. This will be needed for some problems before we can subtract, multiply or divide.

MIXED NUMBERS: Review the meaning of a mixed number.

A mixed number is a number containing a whole number and a fraction.

Look at this subtraction problem: $1\frac{1}{3} - \frac{2}{3} =$

This is pretty simple and most of the kids will tell you that the answer is $\frac{2}{3}$. But we will use this as a demonstration so they will understand what to do with the harder problems.

You can demonstrate this with circles drawn on the board, and if you like, you can give students the three-page set of fractional circles (pages **66-68**) so they can demonstrate this to themselves at their desks. Laminate circles and have students shade in sections with crayon and then clean with facial tissue.

Draw two circles on the board and divide each into thirds. Shade in $1\frac{1}{3}$ of the circles. This makes it easy to see that $1\frac{1}{3} = \frac{4}{3}$.

Now if we subtract $\frac{4}{3} - \frac{2}{3}$ it is easier to see that the answer is $\frac{2}{3}$.

Look at another example: $2\frac{1}{3}$. Again draw circles and demonstrate that $2\frac{1}{3} = \frac{7}{3}$.

When we want to change from a mixed number to an improper fraction without drawing circles, multiply the whole number times the denominator.

For $2\frac{1}{3}$, how many thirds are in the whole number 2? Multiply $2 \times 3 = 6$. Add the other $\frac{1}{3}$ and get $\frac{7}{3}$.

$$2\frac{1}{3} = ? \qquad 2\frac{1}{3} = \frac{7}{3}$$

$$(3 \times 2 = 6, 6 + 1 = 7)$$

Demonstrate one or two more and also get the answer by multiplying and adding. $4\frac{1}{4}$.

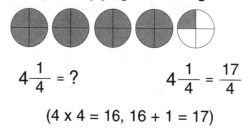

$$4\frac{1}{4} = ? \qquad 4\frac{1}{4} = \frac{17}{4}$$

$$(4 \times 4 = 16, 16 + 1 = 17)$$

How about $5\frac{1}{6}$?

$$5\frac{1}{6} = ? \qquad 5\frac{1}{6} = \frac{31}{6}$$

$$(5 \times 6 = 30, 30 + 1 = 31)$$

RULE: To change from a mixed number to an improper fraction, multiply the denominator times the whole number. Then add the numerator.

$$2\frac{1}{3} = \frac{7}{3}$$

$$(3 \times 2 = 6, 6 + 1 = 7)$$

MIXED NUMBERS

Directions: Change these mixed numbers to improper fractions.

1. $1\frac{2}{3} = \frac{5}{3}$ 2. $2\frac{1}{2} = $ __   3. $2\frac{3}{4} = $ __   4. $2\frac{1}{5} = $ __   5. $3\frac{1}{3} = $ __

$(3 \times 1 = 3, 3 + 2 = 5)$

6. $2\frac{7}{10} = $ __   7. $2\frac{5}{8} = $ __   8. $3\frac{3}{4} = $ __   9. $3\frac{5}{9} = $ __   10. $5\frac{2}{5} = $ __

Directions: Change to improper fractions and subtract. Reduce.

11. $2\frac{2}{5} = $ __   12. $3\frac{5}{8} = $ __   13. $2\frac{1}{10} = $ __   14. $2\frac{1}{7} = $ __   15. $5\frac{1}{5} = $ __

  $- 1\frac{4}{5} = $ __ $- 1\frac{7}{8} = $ __ $- 1\frac{9}{10} = $ __ $- 1\frac{5}{7} = $ __ $- 2\frac{4}{5} = $ __

Directions: Add. For the answer, change the improper fraction to a mixed number. Reduce.

16. $\frac{3}{4} = $ __   17. $\frac{2}{3} = $ __   18. $\frac{9}{10} = $ __   19. $\frac{7}{8} = $ __   20. $\frac{5}{6} = $ __

  $+ \frac{7}{8} = $ __ $+ \frac{5}{6} = $ __ $+ \frac{4}{5} = $ __ $+ \frac{2}{3} = $ __ $+ \frac{11}{12} = $ __

Directions: Add or subtract. When answers are improper fractions, change to mixed numbers. Reduce if necessary.

21. $1\frac{2}{3} = $ __   22. $2\frac{2}{5} = $ __   23. $\frac{11}{12} = $ __   24. $3\frac{1}{6} = $ __   25. $4\frac{1}{3} = $ __

  $+ \frac{2}{3} = $ __ $- \frac{4}{5} = $ __ $+ \frac{3}{4} = $ __ $- \frac{5}{6} = $ __ $- \frac{2}{3} = $ __

CHALLENGE:

Your class trip took $1\frac{1}{2}$ hours travel time, $2\frac{2}{3}$ hours at the science center and $1\frac{1}{4}$ hours for lunch. What was the total time for the trip? _____

Name _____

MIXED NUMBERS

Shade in all the squares that have an improper fraction or a mixed number to find a student's favorite number. Two have been done for you.

$\frac{3}{4}$	$\frac{1}{2}$	$\frac{1}{3}$	$\frac{2}{3}$	$\frac{4}{5}$	$\frac{2}{10}$	$\frac{3}{5}$	$\frac{3}{4}$	$\frac{5}{9}$	$\frac{21}{22}$
$\frac{1}{2}$	$1\frac{1}{2}$	$\frac{1}{4}$	$\frac{5}{3}$	$2\frac{1}{3}$	$\frac{21}{8}$	$\frac{2}{3}$	$1\frac{1}{2}$	$\frac{10}{5}$	$2\frac{1}{2}$
$\frac{1}{3}$	$\frac{7}{3}$	$\frac{1}{5}$	$1\frac{1}{4}$	$\frac{5}{6}$	$\frac{16}{3}$	$\frac{1}{2}$	$2\frac{1}{3}$	$\frac{11}{12}$	$3\frac{1}{3}$
$\frac{1}{5}$	$\frac{3}{2}$	$\frac{1}{6}$	$2\frac{1}{3}$	$\frac{6}{7}$	$\frac{19}{4}$	$\frac{9}{11}$	$4\frac{1}{5}$	$\frac{13}{15}$	$\frac{11}{8}$
$\frac{2}{10}$	$\frac{10}{7}$	$\frac{1}{7}$	$\frac{11}{7}$	$\frac{7}{8}$	$\frac{21}{5}$	$\frac{9}{12}$	$3\frac{1}{8}$	$\frac{16}{20}$	$\frac{12}{5}$
$\frac{3}{4}$	$2\frac{1}{5}$	$\frac{1}{8}$	$1\frac{1}{4}$	$\frac{8}{9}$	$\frac{18}{3}$	$\frac{3}{8}$	$\frac{9}{5}$	$\frac{15}{19}$	$6\frac{1}{3}$
$\frac{7}{8}$	$3\frac{1}{8}$	$\frac{1}{9}$	$\frac{10}{6}$	$\frac{9}{10}$	$\frac{17}{2}$	$\frac{1}{3}$	$\frac{8}{3}$	$\frac{13}{15}$	$3\frac{1}{2}$
$\frac{6}{10}$	$\frac{7}{2}$	$\frac{1}{10}$	$\frac{7}{3}$	$\frac{10}{11}$	$\frac{15}{6}$	$\frac{1}{5}$	$\frac{10}{5}$	$\frac{12}{13}$	$4\frac{1}{8}$
$\frac{3}{5}$	$\frac{11}{8}$	$\frac{11}{12}$	$\frac{8}{5}$	$3\frac{1}{5}$	$\frac{18}{5}$	$\frac{9}{10}$	$\frac{9}{2}$	$\frac{8}{3}$	$3\frac{1}{4}$
$\frac{2}{8}$	$\frac{3}{5}$	$\frac{4}{9}$	$\frac{3}{5}$	$\frac{4}{7}$	$\frac{3}{8}$	$\frac{8}{11}$	$\frac{5}{6}$	$\frac{16}{20}$	$\frac{19}{21}$

MIXED NUMBERS

Directions: Find the equivalent fractions.

1. $\frac{1}{7} = \frac{}{63}$ 2. $\frac{1}{4} = \frac{}{24}$ 3. $\frac{1}{8} = \frac{}{72}$ 4. $\frac{1}{3} = \frac{}{24}$ 5. $\frac{3}{9} = \frac{}{63}$

6. $\frac{7}{10} = \frac{}{80}$ 7. $\frac{1}{2} = \frac{}{14}$ 8. $\frac{5}{7} = \frac{}{49}$ 9. $\frac{3}{8} = \frac{}{56}$ 10. $\frac{7}{10} = \frac{}{50}$

Directions: Reduce to lowest terms.

11. $\frac{14}{16} = $ ___   12. $\frac{15}{18} = $ ___   13. $\frac{9}{12} = $ ___   14. $\frac{21}{35} = $ ___   15. $\frac{24}{36} = $ ___

16. $\frac{27}{30} = $ ___   17. $\frac{35}{56} = $ ___   18. $\frac{56}{64} = $ ___   19. $\frac{72}{80} = $ ___   20. $\frac{81}{90} = $ ___

Directions: Change to improper fractions if you need to.

21. $1\frac{7}{8} = $ ___   22. $2\frac{1}{4} = $ ___   23. $2\frac{1}{5} = $ ___   24. $\frac{5}{6} = $ ___   25. $2\frac{1}{10} = $ ___

$+\ 2\frac{2}{8} = $ ___ $-\ 1\frac{3}{4} = $ ___ $+\ 1\frac{2}{5} = $ ___ $+\ \frac{5}{6} = $ ___ $-\ 1\frac{3}{10} = $ ___

26. The weather bureau says we received $1\frac{1}{10}$ inches of rain on Monday and $1\frac{1}{2}$ inches of rain on Tuesday. What was the total for those two days? _____

27. The gym teacher needs two new jump ropes. Each one is $7\frac{1}{3}$ feet long. How many feet of rope will be needed in all? _____

CHALLENGE:

1. Your mom bought $2\frac{7}{8}$ pounds of chocolate chips. She used $1\frac{1}{2}$ pounds in the cookies. How many pounds are left? _____

2. If you lift weights $1\frac{1}{3}$ hours on Saturday and $1\frac{2}{5}$ hours on Sunday, what is your total for the two days? _____

3. Your neighbor's baby weighed $6\frac{3}{4}$ pounds at birth and has gained $1\frac{1}{3}$ pounds. How much does the baby weigh now? _____

INTRODUCTION: This is a definite shift from the addition and subtraction. There will still be some reducing involved, but this is a very different process from what has been done up to now in fractions. Take a little extra care in this step so that students understand the basic ideas involved.

In Lesson Plan 2 (pages 15-16), students needed to understand the basic idea involved in adding and subtracting fractions–adding and subtracting numerators and keeping the same denominator. In multiplication of fractions, they need to understand the basic ideas–that *times* means "of" and *of* means "times," and that to multiply fractions, you multiply numerators and multiply denominators.

MULTIPLYING FRACTIONS: When we multiply whole numbers–for example, 3 x 4–we are taking three groups of four or four groups of three.

When we multiply fractions–for example, $\frac{1}{3}$ x $\frac{1}{4}$–we are taking $\frac{1}{3}$ of $\frac{1}{4}$, so we say

TIMES (X) MEANS OF.

If we have a word problem telling us $\frac{1}{3}$ of the class is sick and $\frac{1}{4}$ of the sick kids have the flu, to find out what fraction of the kids have the flu, we need to take $\frac{1}{4}$ of $\frac{1}{3}$. We need to multiply those fractions.

OF MEANS TIMES (X).

Please take the time to demonstrate this carefully. Concepts involved in fractions are very simple, but many people cannot remember which steps go with addition and which steps go with multiplication because they did not understand them in the beginning.

Following are several demonstrations. Use as many as you need to get through to the class but not so many as to bore them.

DEMONSTRATION 1: If you cut your candy bar into fourths and then cut each piece into thirds, what fractional part will each piece be?

Use fraction strips and rectangles on the board or overhead projector. Laminated fraction strips work the best. Students can shade in areas and then rub off the shading with facial tissue before the next problem. Have each student find the answer. Start with the strip that has fourths. With dotted lines, cut each fourth into three pieces. Now compare with the strip that is divided into twelfths. This shows the multiplication problem.

$$\frac{1}{3} \text{ of } \frac{1}{4} = \frac{1}{3} \times \frac{1}{4} = \frac{1}{12}$$

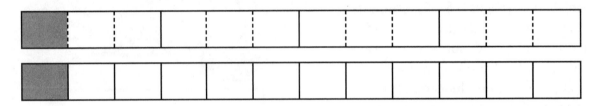

Each piece is $\frac{1}{12}$ of the candy bar. We are really finding $\frac{1}{3}$ of $\frac{1}{4} = \frac{1}{12}$

Rather than draw rectangles, we can multiply numerators and multiply denominators.

$$\frac{1}{3} \times \frac{1}{4} = \frac{1}{12}$$

DEMONSTRATION 2: You can also use fraction strips on the overhead projector or draw fraction strips on the board.

Half of the class is girls. $\frac{2}{3}$ of the girls are taking sewing class. What fractional part of the class is taking sewing? (We will assume that none of the boys will be taking sewing.)

Start with the fraction bar for $\frac{1}{2}$. Mark it into thirds and shade in two of them. Now look at this multiplication problem:

$$\frac{1}{2} \times \frac{2}{3} = \frac{2}{6} = \frac{1}{3}$$

$\frac{2}{6}$ will reduce to $\frac{1}{3}$. Get out the fraction bar for $\frac{1}{3}$ and see that it is the same as the $\frac{2}{3}$ of $\frac{1}{2}$ that we shaded earlier.

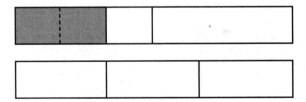

TLC10112 Copyright © Teaching & Learning Company, Carthage, IL 62321-0010

DEMONSTRATION 3: Try 3/4 x 1/2 using fraction strips.

This time let's multiply first and check with the fraction strips.

$$\frac{3}{4} \times \frac{1}{2} = \frac{3}{8}$$

Get out 1/2 fraction strip. Divide 1/2 into four equal parts. Shade in three of the parts. Now compare with the 1/8 fraction strip and see that the answer is 3/8.

You can make up more demonstrations if necessary. Here are some good possibilities: 1/2 x 1/4, 1/2 x 1/16, 3/4 x 1/3, 2/5 x 1/2, 2/3 x 3/4.

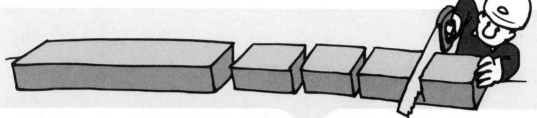

MIXED NUMBERS IN MULTIPLICATION:

The purpose of this *Math Phonics*™ book is to introduce the basic concepts involved in adding, subtracting, multiplying and dividing fractions. For that reason, we will not go into detail about teaching multiplication of mixed numbers. However, if you have the time, here is one you could demonstrate.

$$\frac{1}{3} \times 1\frac{1}{2}$$

$$\frac{1}{3} \times \frac{3}{2} = \frac{3}{6} = \frac{1}{2}$$

Get out the fraction strips for thirds. Shade in one third and half of the next third. Compare with the fraction strip for halves to see that the answer of 1/2 is correct.

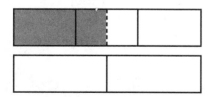

An extra worksheet for multiplying mixed numbers (Level 2) (page 78) is provided if you have time and want to include it in this lesson.

NOTE: Recall that in Lesson Plan 3 students multiplied both numerator and denominator by the same number to find an equivalent fraction.

$$\frac{1}{2} \times \frac{3}{3} = \frac{3}{6}$$

What we are really doing is multiplying

$$\frac{1}{2} \times \frac{3}{3}$$

Since 3/3 = 1, we are multiplying by one and the fraction value stays the same though written with different numbers.

Show your students the connection between these two lessons.

NOTE: The recipe on Worksheet U (page 76) is an actual recipe in case anyone wants to try it at home.

MULTIPLYING FRACTIONS

Directions: Multiply the numerators. Multiply the denominators. Reduce answers.

1. $\frac{1}{3}$ X $\frac{1}{4}$ =

2. $\frac{1}{2}$ X $\frac{1}{6}$ =

3. $\frac{1}{2}$ X $\frac{2}{5}$ =

4. $\frac{2}{3}$ X $\frac{1}{4}$ =

5. $\frac{1}{3}$ X $\frac{6}{8}$ =

6. $\frac{5}{6}$ X $\frac{3}{5}$ =

Review

Change to improper fractions and add. Reduce answers.

7. $2\frac{2}{3}$
$+ \ 1\frac{2}{3}$

8. $2\frac{3}{4}$
$+ \ 1\frac{3}{4}$

9. $1\frac{1}{2}$
$+ \ 2\frac{1}{2}$

10. $1\frac{5}{6}$
$+ \ 2\frac{5}{6}$

11. $1\frac{7}{8}$
$+ \ 3\frac{5}{8}$

Review

Find the common denominator and add. Reduce answers.

12. $\frac{7}{8}$
$+ \ \frac{1}{3}$

13. $\frac{9}{10}$
$+ \ \frac{3}{5}$

14. $\frac{6}{7}$
$+ \ \frac{1}{3}$

15. $\frac{5}{6}$
$+ \ \frac{2}{5}$

16. $\frac{3}{4}$
$+ \ \frac{2}{10}$

Oven Caramel Corn

You are going to make $^1/_2$ of the recipe. (Remember that *of* means "times.")

3³/₄ quarts popped corn $^1/_4$ cup light corn syrup $^1/_2$ teaspoon baking soda

1 cup brown sugar $^1/_2$ teaspoon salt   1 teaspoon vanilla

$^1/_2$ cup butter

Measure brown sugar, margarine, syrup and salt into saucepan. Stir well and boil 5 minutes. Remove from heat and stir in vanilla and soda. Pour over popped corn in large container. Mix well and spread on two large cookie sheets. Bake at 250°F for 1 hour, stirring occasionally. Remove from oven and separate kernels. Add peanuts if desired.

17. How much brown sugar will you need for $^1/_2$ of the recipe? _____

18. How much salt? _____

19. How much light corn syrup? _____

CHALLENGE:

How many quarts of popped corn? _____

MULTIPLYING FRACTIONS

Multiply the fractions in the design. Use these colors for each space with these answers: $1/12$ = red; $1/16$ = blue; $1/24$ = purple; $1/36$ = yellow; $1/25$ = green; $1/35$ = orange; all others, do not color.

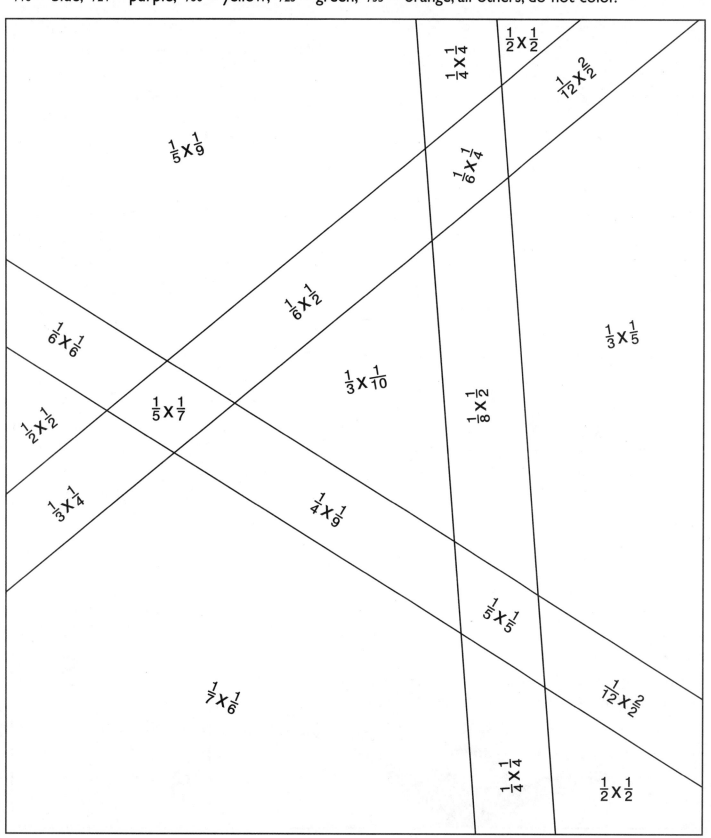

MULTIPLYING FRACTIONS

Multiply. Reduce if possible. Change improper fractions to mixed numbers. Show your work on the back.

1.

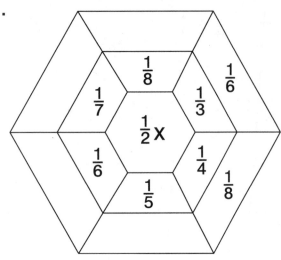

2.

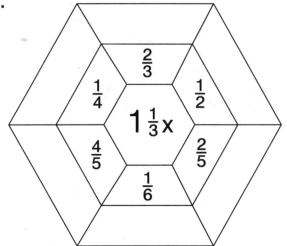

3.

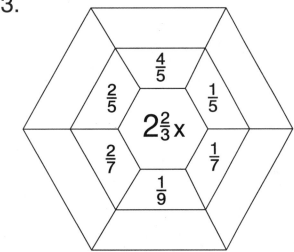

4.

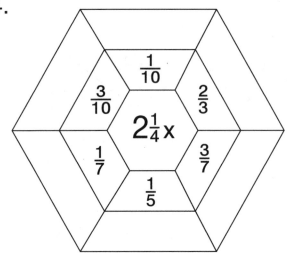

CHALLENGE:

In this magic square, the total for each row—vertical, horizontal or diagonal—is 21. Fill in the missing numbers.

INTRODUCTION: Here we will show that division and multiplication are related to each other in working with fractions. We will also show the shortcut of dividing fractions—inverting the second fraction.

PREPARATION: Before talking about division of fractions, talk about expressing whole numbers as fractions. If you have two cherry pies each cut into thirds, you have $6/3$, and $6/3 = 2$. If you have two pies each cut into halves, you have $4/2$, and $4/2 = 2$. Tell the class that there is a way to write about those pies as a fraction even if they haven't been cut. If you have two pies uncut, you can say you have $2/1$, and $2/1 = 2$.

DIVISION OF FRACTIONS: Now show the class the first example of division of fractions—$2/3 \div 1/3$.

When we say $10 \div 5$, we want to know how many groups of five are contained in the group of 10. The answer is two.

In $2/3 \div 1/3$, we want to know how many times the $1/3$ is contained in the $2/3$. It seems that the answer should be 2. In multiplication of fractions, we multiplied numerators and multiplied denominators. What happens if we divide numerators and divide denominators?

$$\frac{2}{3} \div \frac{1}{3} = \frac{2}{1} = 2$$

We get the answer of $2/1$, and we saw that $2/1 = 2$.

This is fine as long as numerators and denominators are divisible, but sometimes they are not. What if we had $5/6 \div 3/4$? Another way to think of multiplication and division is that they are opposites or inverses of each other.

For fractions, an inverse can be made by switching numerators and denominators.

The inverse of $3/4$ is $4/3$. The inverse of $5/6$ is $6/5$.

So, to divide, we can change the problem to

$$\frac{2}{3} \times \frac{3}{1} = \frac{6}{3} = 2$$

This way we also get the same answer of 2.

Let's try $3/4 \div 1/4$. It seems that the answer should be 3.

Try dividing numerators and dividing denominators.

$$\frac{3}{4} \div \frac{1}{4} = \left(\begin{matrix} 3 \div 1 = 3 \\ 4 \div 4 = 1 \end{matrix}\right) \frac{3}{1} = 3$$

Try using the inverse.

$$\frac{3}{4} \times \frac{4}{1} = \frac{12}{4} = 3$$

We will use the inverse on the worksheets.

LEVEL 2: Show students how to factor a number to primes.

DIVISION OF FRACTIONS

Directions: Change these mixed numbers by dividing. Reduce.

$\frac{27}{6} = 4\frac{1}{2}$ 1. $\frac{29}{7} =$ ___   2. $\frac{32}{5} =$ ___   3. $\frac{66}{9} =$ ___   4. $\frac{25}{8} =$ ___

$$6\overline{)27} \quad \begin{array}{c} 4\frac{3}{6} = 4\frac{1}{2} \\ \underline{24} \\ 3 \end{array}$$

Directions: Change to improper fractions.

$4\frac{1}{3} = \frac{13}{4}$ 5. $5\frac{1}{2} =$ ___   6. $6\frac{2}{3} =$ ___   7. $4\frac{1}{4} =$ ___   8. $9\frac{1}{2} =$ ___

$\begin{pmatrix} 3 \times 4 = 12 \\ 12 + 1 = 13 \end{pmatrix}$

Directions: Add or subtract. Reduce answers if possible.

9. $\frac{3}{4} =$ —   10. $2\frac{1}{5} =$ —   11. $\frac{9}{10} =$ —   12. $2\frac{5}{6} =$ —   13. $\frac{7}{8} =$ —

$+ \frac{2}{3} =$ — $- \frac{3}{5} =$ — $- \frac{2}{5} =$ — $- 1\frac{1}{6} =$ — $+ \frac{2}{3} =$ —

Directions: Multiply. Reduce if possible.

14. $\frac{2}{3} \times \frac{1}{4} =$ 15. $\frac{9}{10} \times \frac{2}{3} =$ 16. $\frac{3}{4} \times \frac{1}{2} =$

17. $\frac{5}{6} \times \frac{3}{4} =$ 18. $\frac{4}{5} \times \frac{1}{2} =$ 19. $\frac{7}{8} \times \frac{1}{3} =$

Directions: To divide, invert the second fraction and multiply. Reduce if possible.

20. $\frac{2}{3} \div \frac{1}{3} =$ 21. $\frac{3}{4} \div \frac{1}{4} =$ 22. $\frac{9}{10} \div \frac{1}{10} =$

23. $1\frac{1}{3} \div \frac{2}{3} =$ 24. $1\frac{3}{5} \div \frac{4}{5} =$ 25. $1\frac{3}{7} \div \frac{7}{10} =$

DIVISION OF FRACTIONS

Directions: Multiply. Reduce if you can.

1. $\frac{2}{3} \times \frac{1}{4} =$ 2. $\frac{2}{5} \times \frac{3}{4} =$ 3. $\frac{1}{2} \times \frac{5}{6} =$ 4. $\frac{7}{8} \times \frac{2}{3} =$

Directions: Divide. Reduce if you can. Change improper fractions to mixed numbers.

5. $\frac{3}{4} \div \frac{1}{2} =$ 6. $\frac{7}{8} \div \frac{1}{4} =$ 7. $\frac{9}{10} \div \frac{1}{5} =$ 8. $\frac{5}{6} \div \frac{1}{5} =$

PERFECT NUMBERS: Everybody wants to be perfect every now and then. Some numbers are perfect. A perfect number is equal to the sum of all its smaller factors. Six is the smallest perfect number. It is divisible by 1, 2 and 3. 1 + 2 + 3 = 6. Do you know what the next perfect number is? Do these problems and fill in the puzzle below to find out.

Reduce.

$\frac{9}{12} = \frac{3}{4}$ **E**	$\frac{8}{10} = -$ **T**	$\frac{25}{35} = -$ **N**	$\frac{63}{70} = -$ **I**
$\frac{6}{12} = -$ **H**	$\frac{10}{24} = -$ **W**	$\frac{18}{21} = -$ **Y**	$\frac{9}{27} = -$ **G**

$\underline{}$ $\underline{}$ $\underline{E}$ $\underline{}$ $\underline{}$ $\underline{}$ $\overline{}$ $\underline{E}$ $\underline{}$ $\underline{}$ $\underline{}$ $\underline{}$
$\frac{4}{5}$ $\frac{5}{12}$ $\frac{3}{4}$ $\frac{5}{7}$ $\frac{4}{5}$ $\frac{6}{7}$ $\frac{3}{4}$ $\frac{9}{10}$ $\frac{1}{3}$ $\frac{1}{2}$ $\frac{4}{5}$

Can you think of all the smaller factors of this number? Do they add up to this number?

DIVISION OF FRACTIONS

Directions: Change to improper fractions and multiply. Reduce answers.

1. $1\frac{1}{2} \times 2\frac{1}{3} =$

2. $2\frac{7}{10} \times \frac{2}{3} =$

3. $1\frac{3}{4} \times \frac{1}{7} =$

4. $2\frac{1}{2} \times 1\frac{1}{2} =$

5. $3\frac{1}{3} \times \frac{3}{10} =$

6. $5\frac{1}{2} \times \frac{2}{11} =$

7. $6\frac{1}{3} \times \frac{3}{4} =$

8. $3\frac{6}{10} \times \frac{2}{3} =$

Directions: Reduce answers.

9. $1\frac{1}{2} \div \frac{2}{3} =$

10. $2\frac{1}{3} \div \frac{1}{3} =$

11. $4\frac{1}{2} \div \frac{2}{9} =$

12. $1\frac{1}{5} \div 2\frac{2}{5} =$

13. $3\frac{5}{10} \div \frac{5}{10} =$

14. $6\frac{3}{7} \div \frac{5}{7} =$

15. $3\frac{1}{4} \div 2\frac{1}{2} =$

16. $1\frac{1}{6} \div 2\frac{1}{2} =$

Directions: Find the prime factors.

17. $12 = 3 \times 4 = 3 \times 2 \times 2$

18. $27 =$

19. $54 =$

20. $72 =$

CHALLENGE:

An 1850s log cabin had these dimensions: width = $10\frac{1}{2}$ feet, length = $11\frac{2}{3}$ feet, height = $8\frac{3}{4}$ feet. You are making a model $\frac{1}{7}$ the actual size. What are the dimensions of your model? Show your work on the back.

LESSON PLAN 10

In Lesson Plan 5 (pages 53-54) we learned two rules for finding common denominators. They were:

1. If the larger denominator is a multiple of the smaller one, use the larger as the common denominator. (Ex: $1/2 + 1/4 - 4$ is the common denominator.)

2. Multiply the two denominators to find a common denominator. (Ex: $1/4 + 1/5 - 20$ is the common denominator.) This rule always works, but for larger numbers, it is easier to make a mistake.

When larger numbers are involved, sometimes it is good to use a third method. This method involves primes.

PREPARATION: If your class has not studied primes in the past, you should teach them before teaching this way of finding common denominators.

PRIMES: A prime is a number which has exactly two factors—itself and one. One is not a prime because it has only one factor. Two is a prime; three is a prime. Four is not a prime because it has three factors—1, 2 and 4.

Math Phonics™–*Multiplication* and *Math Phonics*™–*Division* both have activity sheets which demonstrate primes to students. If you do not have those books, use this activity with your class.

DISCOVERING PRIMES: Have each student write all numbers 1-100 on a sheet of notebook paper.

1. Since 2 is a prime, they circle 2 and cross off 4, 6, 8 and all other multiples of 2 up to 100. None of them can be primes because they have at least three factors–themselves, 1 and 2.

2. 3 is a prime, so circle 3. Cross off all multiples of 3.

3. 4 was crossed off because it is a multiple of 2.

4. 5 has not been crossed off. It is a prime, so circle the 5. Now cross off all multiples of 5. Some of the multiples of 5 have already been crossed off–10 and 20 for example. Ask the class why they have already been crossed off. It is because they are also multiples of 2.

5. 6 has been crossed off. 7 has not been crossed off, so it should be circled–it is a prime. Now cross off all multiples of 7. Quiz the class as to why some of these numbers have already been crossed off–why has 35 been crossed off? It is because it is also a multiple of 5.

6. What is the next number that has not been crossed off? It is 11. Circle the 11. Check all the multiples of 11. Have all of them already been crossed off? Why? Go through the multiples of 11 and have the class tell why they have already been crossed off. 22 is also a multiple of 2, 33 is also a multiple of 3 and so on.

7. What is the next prime? 13. Do we need to check the multiples of 13? We do not because all have been crossed off because they were multiples of one of the smaller primes.

Now list all the primes from 1-100. 2, 3, 5, 7, 11, 13, 17, 19, 23, 29, 31, 37, 41, 43, 47, 53, 59, 61, 67, 71, 73, 79, 83, 89, 97.

The primes that will be used most are 2, 3, 5, 7 and 11.

Primes are not used all the time in upper level math courses, but at times they are very important. Some examples are finding common denominators and solving quadratic equations in algebra. It is not essential that you go through this exercise with your students, but if they are made familiar with primes in the lower grades, certain problems are much easier later on.

LEAST COMMON DENOMINATORS:

Now back to the common denominators.

Let's say we had a problem of $1/8 + 1/12$. 12 is not a multiple of 8, so we could multiply the two numbers together and get a common denominator of 96. ($8 \times 12 = 96$) It's better to use a smaller number because when working with larger numbers, we are more likely to make a mistake.

If students are familiar with multiples of 8 and multiples of 12, they might think of 24 as being in both lists and 24 would work. Here is a way to figure out the 24 if you could not think of the multiples.

We will write 8 and 12 as factors which are all primes.

$$8 = 2 \times 4 = 2 \times 2 \times 2$$
$$12 = 3 \times 4 = 3 \times 2 \times 2$$

The common denominator should have prime factors which will cover the prime factors in the 8 and the 12.

The common denominator is $2 \times 2 \times 2 \times 3 = 24$.

We don't need any more 2s for the 2s in the factors for 12 because they are already covered by the 2s in the 8.

Here is another example. $5/6 + 7/15 =$

$6 \times 15 = 90$, so we could use 90 as the common denominator. But we could find a smaller denominator by using primes.

$$6 = 3 \times 2$$
$$15 = 5 \times 3$$

The least common denominator is $5 \times 3 \times 2 = 30$.

This keeps the numbers smaller and easier to work with.

LEAST COMMON DENOMINATORS

Remember these rules of divisibility:

Divisible by 2—ends in 0, 2, 4, 6 or 8.

Divisible by 3—numerals add up to 3, 6 or 9.

Divisible by 5—ends in 0 or 5.

Directions: Circle the primes. Cross out all others.

1.   2, 3, 4, 5, 6, 7, 8, 9, 10, 11, 12, 13

Directions: Factor these numbers until you have all primes.

2. 6 = __ x __

3. 8 = __ x __ = __ x __ x __

4. 12 = __ x __ = __ x __ x __

5. 18 = __ x __ = __ x __ x __

6. 25 = __ x __

7. 36 = __ x __ = __ x __ x __ = __ x __ x __ x __

Directions: Factor both numbers. Find the least common denominator.

8. 6 = __ x __

8 = __ x __ x __

Least common denominator is

__ x __ x __ x __ = __

9. 12 = __ x __ = __ x __ x __

9 = __ x __

Least common denominator is

__ x __ x __ x __ = __

10. 10 = __ x __

6 = __ x __

Least common denominator is

__ x __ x __ = __

11. 10 = __ x __

8 = __ x __ x __

Least common denominator is

__ x __ x __ x __ = __

CHALLENGE:

Factor all three numbers. Find the least common denominator. If the number is already a prime, write *prime* after the equal sign.

1. 3 =

6 =

10 =

L. C. D. =

2. 8 =

10 =

5 =

L. C. D. =

3. 15 =

10 =

6 =

L. C. D. =

4. 18 =

10 =

3 =

L. C. D. =

LEAST COMMON DENOMINATOR

Directions: Add or subtract. Use prime factors if you need to for finding the least common denominators. Reduce if possible.

1. $\frac{1}{4}$ = $\frac{3}{12}$

+ $\frac{1}{6}$ = $\frac{2}{12}$

$\frac{5}{12}$

4 = 2 x 2
6 = 3 x 2
L. C. D. = 3 x 2 x 2 = 12

2. $\frac{1}{9}$ = ___

+ $\frac{1}{6}$ = ___

9 =
6 =
L. C. D. =

3. $\frac{1}{4}$ = ___

+ $\frac{1}{12}$ = ___

4 =
12 =
L. C. D. =

4. $\frac{1}{4}$ = ___

$\frac{1}{3}$ = ___

+ $\frac{1}{6}$ = ___

4 =
3 =
6 =
L. C. D. =

5. $\frac{1}{15}$ = ___

+ $\frac{1}{9}$ = ___

15 =
9 =
L. C. D. =

6. $\frac{1}{25}$ = ___

+ $\frac{1}{10}$ = ___

25 =
10 =
L. C. D. =

7. $\frac{1}{4}$ = ___

+ $\frac{1}{9}$ = ___

4 =
9 =
L. C. D. =

8. $\frac{1}{14}$ = ___

+ $\frac{1}{10}$ = ___

14 =
10 =
L. C. D. =

9. $\frac{1}{16}$ = ___

+ $\frac{1}{5}$ = ___

16 =
5 =
L. C. D. =

CHALLENGE:

1. $2\frac{1}{8}$ = ___

$3\frac{1}{3}$ = ___

+ $1\frac{1}{9}$ = ___

8 =
3 =
9 =
L. C. D. =

2. $\frac{1}{15}$ = ___

$\frac{1}{5}$ = ___

+ $\frac{1}{9}$ = ___

15 =
5 =
9 =
L. C. D. =

Name _____

LEAST COMMON DENOMINATORS

Directions: Reduce answers to lowest terms. Change all improper fractions to mixed numbers.

1. $10\frac{1}{3}$
    $-\ \ 7\frac{1}{4}$

2. $6\frac{1}{8}$
    $-\ \ 4\frac{5}{6}$

3. $4\frac{3}{4}$
    $10\frac{1}{7}$
    $+\ \ 5\frac{1}{2}$

4. $18\frac{1}{2}$
    $21\frac{9}{10}$
    $+\ \ 13\frac{1}{4}$

5. $16\frac{2}{3}$
    $-\ 10\frac{6}{7}$

6. $19\frac{3}{25}$
    $25\frac{1}{2}$
    $+\ \ 13\frac{9}{10}$

7. $35\frac{1}{7}$
    $-\ 29\frac{4}{5}$

8. $16\frac{1}{4}$
    $18\frac{7}{8}$
    $+\ \ 23\frac{1}{3}$

9. $36\frac{2}{3}$
    $-\ 19\frac{9}{10}$

10. $\frac{3}{14}$
     $\frac{5}{10}$
     $+\ \ \frac{2}{35}$

11. $\frac{1}{2}$
     $\frac{5}{6}$
     $+\ \ \frac{3}{14}$

12. $\frac{4}{15}$
     $\frac{3}{10}$
     $+\ \ \frac{5}{6}$

CHALLENGE:

Here is a recipe for trail mix.

  $2^{1}/_{12}$ cups raisins
  $1^{2}/_{3}$ cups chocolate chips
  $3^{1}/_{4}$ cups peanuts

If you want to make three recipes, how many cups of each item do you need? _____
How many cups of trail mix will you have from the three recipes? _____ If you have 14
students in your class and you make three recipes of trail mix, how many cups of mix will each student get? _____

RULES

FRACTION: A number that shows that something has been divided. It has the numerator above the line and the denominator below the line.

ADDITION OF FRACTIONS: Add the numerators. Keep the same denominator. $\frac{1}{3} + \frac{1}{3} = \frac{2}{3}$

SUBTRACTION OF FRACTIONS: Subtract the numerators. Keep the same denominator. $\frac{4}{5} - \frac{1}{5} = \frac{3}{5}$

EQUIVALENT FRACTIONS: Two fractions with the same value written with different numbers. $\frac{1}{4} = \frac{3}{12}$

ZEROS IN DIVISION: Zero cannot be the divisor. Zero cannot be the denominator. You cannot divide by zero.

COMMON DENOMINATORS:

1. See if the larger denominator is a multiple of the smaller one. If so, use the larger one as a common denominator. $\frac{1}{2} = \frac{2}{4}$

$$+ \ \frac{1}{4} = \frac{1}{4}$$
$$\overline{\phantom{+ \ \frac{1}{4} = }\frac{3}{4}}$$

2. Multiply the two denominators to get a common denominator. (This one always works.) $\frac{1}{4} = \frac{5}{20}$

$$+ \ \frac{1}{5} = \frac{4}{20}$$
$$\overline{\phantom{+ \ \frac{1}{5} = }\frac{9}{20}}$$

3. Write each denominator as the product of primes. Use the primes to form the common denominator.

IMPROPER FRACTIONS: An improper fraction has a numerator larger than the denominator. $\frac{7}{6}$

PROPER FRACTIONS: A proper fraction has a numerator smaller than the denominator. $\frac{3}{4}$

MIXED NUMBERS: Mixed numbers have a whole number and a fraction. $6\frac{3}{4}$

MULTIPLYING FRACTIONS: Multiply numerators; multiply denominators. $\frac{2}{3} \times \frac{4}{5} = \frac{8}{15}$

DIVIDING FRACTIONS: Invert the divisor and multiply. $\frac{3}{4} \div \frac{1}{4} = \frac{3}{4} \times \frac{4}{1} = \frac{12}{4} = 3$

GAMES

Name _____

FRACTION ASSESSMENT

A. Reduce.

$\frac{4}{10} =$ $\frac{6}{9} =$ $\frac{7}{21} =$ $\frac{25}{40} =$ $\frac{10}{50} =$

B. Make equivalent fractions.

$\frac{1}{2} = \frac{}{10}$ $\frac{2}{5} = \frac{}{15}$ $\frac{3}{4} = \frac{}{12}$ $\frac{9}{10} = \frac{}{80}$ $\frac{7}{8} = \frac{}{64}$

C. Change to mixed numbers.

$\frac{5}{4} =$ $\frac{9}{7} =$ $\frac{21}{12} =$ $\frac{17}{4} =$ $\frac{24}{11} =$

D. Change to improper fractions.

$2\frac{1}{2} =$ $3\frac{3}{4} =$ $5\frac{2}{3} =$ $4\frac{5}{6} =$ $7\frac{3}{5} =$

Directions: Reduce if possible. Change improper fractions to mixed numbers.

E. Make equivalent fractions.

$\begin{array}{r} \frac{1}{3} \\ + \frac{1}{3} \\ \hline \end{array}$ $\begin{array}{r} \frac{7}{10} \\ + \frac{2}{10} \\ \hline \end{array}$ $\begin{array}{r} \frac{3}{4} \\ + \frac{3}{4} \\ \hline \end{array}$ $\begin{array}{r} \frac{1}{3} \\ + \frac{1}{6} \\ \hline \end{array}$ $\begin{array}{r} 3\frac{1}{4} \\ + 2\frac{5}{8} \\ \hline \end{array}$

F. Subtract.

$\begin{array}{r} \frac{2}{3} \\ - \frac{1}{3} \\ \hline \end{array}$ $\begin{array}{r} \frac{1}{2} \\ - \frac{1}{4} \\ \hline \end{array}$ $\begin{array}{r} \frac{3}{8} \\ - \frac{1}{4} \\ \hline \end{array}$ $\begin{array}{r} \frac{5}{6} \\ - \frac{1}{3} \\ \hline \end{array}$ $\begin{array}{r} \frac{9}{10} \\ - \frac{1}{2} \\ \hline \end{array}$

G. Multiply.

$\frac{1}{2} \times \frac{1}{3} =$ $\frac{1}{4} \times \frac{2}{5} =$ $\frac{2}{3} \times \frac{4}{5} =$ $\frac{5}{6} \times \frac{3}{4} =$

H. Divide.

$\frac{1}{2} \div \frac{1}{4} =$ $\frac{2}{3} \div \frac{1}{3} =$ $1\frac{1}{3} \div \frac{1}{3} =$ $\frac{9}{10} \div \frac{1}{2} =$

Name _____

A. Reduce.

$\frac{35}{40} =$ $\frac{50}{600} =$ $\frac{28}{36} =$ $\frac{210}{630} =$

B. Make equivalent fractions.

$\frac{5}{9} = \frac{}{54}$ $\frac{7}{8} = \frac{}{72}$ $\frac{5}{6} = \frac{}{42}$ $\frac{6}{9} = \frac{}{81}$

C. Change to mixed numbers.

$\frac{36}{7} =$ $\frac{48}{15} =$ $\frac{250}{75} =$ $\frac{75}{18} =$

D. Change to improper fractions.

$15\frac{2}{3} =$ $21\frac{5}{8} =$ $37\frac{3}{4} =$ $19\frac{3}{8} =$

Directions: Reduce if possible. Change improper fractions to mixed numbers.

E. Add.

$\frac{18}{30}$ $\frac{24}{50}$ $\frac{15}{21}$ $\frac{7}{10}$ $3\frac{1}{5}$

$+ \frac{9}{30}$ $+ \frac{19}{50}$ $+ \frac{6}{7}$ $+ \frac{12}{25}$ $+ 4\frac{1}{3}$

F. Subtract.

$\frac{25}{30}$ $\frac{41}{60}$ $13\frac{1}{8}$ $9\frac{1}{6}$ $12\frac{1}{10}$

$- \frac{18}{30}$ $- \frac{27}{60}$ $- 11\frac{1}{3}$ $- 5\frac{4}{9}$ $- 3\frac{1}{4}$

G. Multiply.

$\frac{5}{8} \times \frac{2}{3} =$ $1\frac{1}{2} \times \frac{3}{4} =$ $3\frac{2}{5} \times \frac{5}{8} =$

H. Divide.

$2\frac{1}{3} \div \frac{1}{3} =$ $3\frac{4}{5} \div 1\frac{1}{3} =$ $3\frac{2}{3} \div 1\frac{1}{2} =$

I. Find prime factors.

$64 =$ $75 =$ $98 =$

ANSWER KEY

Worksheet A, page 12

1. a.   b.   c.

2. $\frac{2}{4}$

3. a. $\frac{1}{2}$ ⟨$\frac{1}{4}$⟩ ⟨$\frac{3}{4}$⟩   b. ⟨$\frac{9}{10}$⟩ ⟨$\frac{2}{10}$⟩ $\frac{2}{5}$
   c. ⟨$\frac{2}{3}$⟩ $\frac{2}{5}$ ⟨$\frac{1}{3}$⟩   d. ⟨$\frac{1}{5}$⟩ ⟨$\frac{3}{5}$⟩ $\frac{3}{4}$

4. $\frac{4}{7}$

5. a. $\frac{2}{9}$ b. $\frac{4}{9}$ c. $\frac{3}{9}$

6. Answers will vary. Some answers might include: recipes, serving, woodworking, wallpaper, wrenches, etc.

Challenge:

1. a. $\frac{5}{9}$ b. $\frac{7}{9}$ c. $\frac{6}{9}$ 2.

3. $\frac{1}{12}$

Worksheet B, page 14

1. a.
   b.
   c.
   d.

2. These fractions should be circled: $\frac{1}{5}$ $\frac{3}{5}$ $\frac{5}{5}$ $\frac{4}{5}$

3. a. $\frac{5}{6}$ b. $\frac{3}{4}$ c. $\frac{9}{10}$ d. $\frac{1}{5}$

4. a. $\frac{4}{7}$ b. $\frac{3}{7}$

5. numerator, denominator

Challenge: 1. Answers will vary. 2. $\frac{5}{7}$, 3. $\frac{4}{15}$, $\frac{5}{15}$

Note: Up until step 4, students are not taught to reduce. $\frac{2}{4}$ is right, but if a student figures out how to reduce, *do not* count the reduced fraction wrong.

Worksheet C, page 23

1. $\frac{2}{4}$ 2. $\frac{4}{5}$ 3. $\frac{8}{11}$ 4. $\frac{7}{10}$ 5. $\frac{16}{20}$ 6. $\frac{14}{15}$

7. $\frac{2}{5}$ 8. $\frac{6}{15}$ 9. $\frac{6}{24}$ 10. $\frac{5}{20}$ 11. $\frac{7}{18}$

12. $\frac{5}{21}$ 13. $\frac{15}{18}$ 14. $\frac{10}{13}$ 15. $\frac{9}{22}$ 16. $\frac{7}{13}$

17. $\frac{11}{20}$ 18. $\frac{13}{14}$ 19. $\frac{7}{10}$ of a mile   20. $\frac{1}{5}$

21. $\frac{4}{6}$ of the money

Challenge: $\frac{30}{100}$, $.30 or 30¢

Worksheet D, page 24

1. $\frac{5}{7}$ 2. $\frac{5}{8}$ 3. $\frac{6}{10}$ 4. $\frac{13}{15}$ 5. $\frac{15}{17}$ 6. $\frac{12}{13}$

7. $\frac{2}{10}$ 8. $\frac{5}{15}$ 9. $\frac{4}{10}$ 10. $\frac{7}{16}$ 11. $\frac{6}{12}$

12. $\frac{5}{13}$ 13. $\frac{6}{7}$ of a gallon   14. $\frac{3}{10}$

D	E	N	O	M	I	N	A	T	O	R
I	S	U	M	A	Y	U	N	S	E	E
V	T	M	P	N	A	M	D	U	D	D
I	A	R	R	I	D	E	R	B	U	U
D	T	T	O	L	D	R	A	T	C	C
E	E	O	D	R	I	A	S	R	E	E
	A	A	U	S	T	T	N	A	L	L
F	R	A	C	T	I	O	N	C	I	I
N	E	S	T	G	O	R	O	T	P	O
G	D	P	A	I	N	U	T	S	L	N

Challenge: in, at, or, may, see, ride, rider, told, rust, rat, nail, act, on, nest, go, rot, pain, nut, nuts, ate, estate, state, rasp, man, add, it, and, not, as, action, ear, are, lion

Worksheet E, page 25

1. $\frac{13}{20}$ 2. $\frac{29}{30}$ 3. $\frac{33}{40}$ 4. $\frac{42}{50}$ 5. $\frac{51}{60}$ 6. $\frac{46}{60}$

7. $\frac{8}{30}$ 8. $\frac{17}{40}$ 9. $\frac{9}{30}$ 10. $\frac{21}{40}$ 11. $\frac{3}{50}$

12. $\frac{7}{30}$ 13. $\frac{17}{20}$ 14. $\frac{7}{20}$ 15. $\frac{14}{30}$ 16. $\frac{43}{60}$

17. $\frac{31}{50}$ 18. $\frac{13}{50}$ 19. $\frac{22}{35}$ 20. $\frac{16}{18}$ 21. $\frac{6}{10}$

Challenge:

$\frac{6}{15}$	$\frac{1}{15}$	$\frac{8}{15}$
$\frac{7}{15}$	$\frac{5}{15}$	$\frac{3}{15}$
$\frac{2}{15}$	$\frac{9}{15}$	$\frac{4}{15}$

Worksheet F, page 35

1. $\frac{3}{6}$ 2. $\frac{2}{8}$ 3. $\frac{6}{9}$ 4. $\frac{10}{25}$ 5. $\frac{8}{12}$ 6. $\frac{18}{24}$

7. $\frac{15}{18}$ 8. $\frac{27}{30}$ 9. $\frac{4}{16}$ 10. $\frac{12}{16}$ 11. $\frac{8}{12}$

12. $\frac{8}{10}$ 13. $\frac{3}{21}$ 14. $\frac{10}{25}$ 15. $\frac{21}{24}$ 16. $\frac{45}{50}$

17. $\frac{4}{6}$ 18. $\frac{4}{5}$ 19. $\frac{3}{12}$ 20. $\frac{7}{15}$ 21. $\frac{13}{25}$

22. $\frac{7}{30}$ 23. $\frac{8}{12}$ of a foot   24. $\frac{55}{100}$

Challenge: $\frac{6}{12}$ of the girls, $\frac{6}{25}$ of the class, $\frac{0}{13}$ of the boys

ANSWER KEY

Worksheet G, page 36

3s: 3, 6, 9, 12, 15, 18, 21, 24, 27, 30

4s: 4, 8, 12, 16, 20, 24, 28, 32, 36, 40

6s: 6, 12, 18, 24, 30, 36, 42, 48, 54, 60

7s: 7, 14, 21, 28, 35, 42, 49, 56, 63, 70

8s: 8, 16, 24, 32, 40, 48, 56, 64, 72, 80

9s: 9, 18, 27, 36, 45, 54, 63, 72, 81, 90

1. $\frac{1}{3}$ 2. $\frac{2}{5}$ 3. $\frac{2}{6}$ 4. $\frac{3}{8}$ 5. $\frac{5}{12}$

6. $\frac{8}{10}$ 7. $\frac{1}{3} > \frac{1}{5}$ 8. $\frac{1}{4} > \frac{1}{6}$ 9. $\frac{3}{8} < \frac{2}{5}$

10. $\frac{5}{12} > \frac{3}{8}$ 11. $\frac{9}{12} = \frac{3}{4}$ 12. $\frac{2}{3} < \frac{7}{10}$

13. $\frac{5}{20}$ 14. $\frac{9}{27}$ 15. $\frac{7}{35}$ 16. $\frac{30}{100}$

17. $\frac{18}{24}$ 18. $\frac{24}{30}$ 19. $\frac{56}{80}$ 20. $\frac{21}{56}$ 21. $\frac{3}{4}$

22. You have more.

Challenge: $\frac{5}{12}$ is left.

Worksheet H, page 37

3s: 3, 6, 9, 12, 15, 18, 21, 24, 27, 30

6s: 6, 12, 18, 24, 30, 36, 42, 48, 54, 60

7s: 7, 14, 21, 28, 35, 42, 49, 56, 63, 70

8s: 8, 16, 24, 32, 40, 48, 56, 64, 72, 80

1. $\frac{35}{42}$ 2. $\frac{63}{72}$ 3. $\frac{35}{49}$ 4. $\frac{72}{81}$ 5. $\frac{48}{56}$ 6. $\frac{45}{55}$

7. $\frac{18}{24}$ 8. $\frac{49}{56}$ 9. $\frac{63}{70}$ 10. $\frac{35}{42}$ 11. $\frac{54}{63}$

12. $\frac{30}{54}$ 13. $\frac{8}{27}$ 14. $\frac{14}{24}$ 15. $\frac{7}{30}$ 16. $\frac{18}{25}$

17. $\frac{31}{40}$ 18. $\frac{8}{15}$ of the students are in their seats.

19. $\frac{92}{100}$ of a dollar

Challenge: You have 5¢ more.

Note: Starting with Worksheet I, all answers should be reduced.

Worksheet I, page 50

1. $\frac{1}{2}$ 2. $\frac{2}{3}$ 3. $\frac{1}{4}$ 4. $\frac{4}{5}$ 5. $\frac{2}{7}$ 6. $\frac{1}{2}$

7. $\frac{3}{5}$ 8. $\frac{1}{3}$ 9. $\frac{4}{9}$ 10. $\frac{2}{3}$ 11. $\frac{1}{4}$ 12. $\frac{7}{8}$

13. $\frac{3}{5}$ 14. $\frac{2}{3}$ 15. $\frac{2}{3}$ 16. $\frac{4}{12} = \frac{1}{3}$

17. $\frac{4}{6} = \frac{2}{3}$ 18. $\frac{5}{10} = \frac{1}{2}$ 19. $\frac{2}{4} = \frac{1}{2}$

20. $\frac{6}{15} = \frac{2}{5}$ 21. $\frac{3}{6} = \frac{1}{2}$ of the pizza

22. $\frac{5}{5}$ of the class

Challenge: $\frac{12}{4} = 3$ dollars, $\frac{12}{3} = 4$ dollars, $\frac{12}{3}$ is the better price.

Worksheet J, page 51

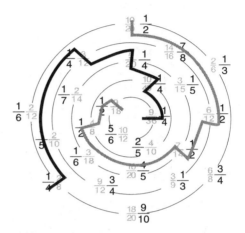

Worksheet K, page 52

1. $\frac{1}{2}$ 2. $\frac{5}{8}$ 3. $\frac{1}{4}$ 4. $\frac{7}{8}$ 5. $\frac{9}{10}$ 6. $\frac{2}{3}$

7. $\frac{9}{10}$ 8. $\frac{8}{9}$ 9. $\frac{7}{9}$ 10. $\frac{4}{5}$ 11. $\frac{3}{5}$ 12. $\frac{3}{5}$

13. $\frac{3}{8}$ 14. $\frac{4}{9}$ 15. $\frac{9}{10}$ 16. $\frac{12}{15} = \frac{4}{5}$

17. $\frac{4}{14} = \frac{2}{7}$ 18. $\frac{32}{33}$ 19. $\frac{30}{50} = \frac{3}{5}$

20. $\frac{42}{50} = \frac{21}{25}$ 21. $\frac{4}{10} = \frac{2}{5}$ of a yard

22. $\frac{33}{60} = \frac{11}{20}$ of an hour

Challenge: Good job!

Worksheet L, page 57

1. $\frac{3}{4}$ 2. $\frac{13}{8}$ 3. $\frac{4}{6} = \frac{2}{3}$ 4. $\frac{5}{10} = \frac{1}{2}$ 5. $\frac{5}{6}$

6. $\frac{9}{10}$ 7. $\frac{5}{16}$ 8. $\frac{10}{12} = \frac{5}{6}$ 9. $\frac{3}{8}$ 10. $\frac{3}{8}$

11. $\frac{3}{6} = \frac{1}{3}$ 12. $\frac{5}{10} = \frac{1}{2}$ 13. $\frac{7}{12}$ 14. $\frac{3}{6} = \frac{1}{2}$

15. $\frac{3}{16}$ 16. $\frac{13}{25}$ 17. $\frac{7}{12}$ 18. $\frac{1}{12}$

19. $\frac{3}{20}$ 20. $\frac{7}{15}$ 21. $\frac{7}{6}$ 22. $\frac{11}{20}$ 23. $\frac{11}{30}$

24. $\frac{13}{42}$

Challenge: $\frac{17}{12}$ cups or $1\frac{5}{12}$ cups

ANSWER KEY

Worksheet M, page 58

1. $\frac{12}{10}$ 2. $\frac{15}{18}$ 3. $\frac{7}{8}$ 4. $\frac{5}{6}$ 5. $\frac{9}{14}$ 6. $\frac{13}{16}$

7. $\frac{27}{30}$ 8. $\frac{17}{22}$ 9. $\frac{1}{4}$ 10. $\frac{11}{21}$ 11. $\frac{5}{12}$

12. $\frac{1}{15}$ 13. $\frac{2}{18} = \frac{1}{9}$ 14. $\frac{7}{36}$ 15. $\frac{1}{18}$

16. $\frac{39}{72} = \frac{13}{24}$ 17. $\frac{16}{21}$ 18. $\frac{5}{12}$ 19. $\frac{11}{36}$

20. $\frac{7}{8}$ 21. $\frac{19}{30}$ 22. $\frac{3}{10}$ 23. $\frac{13}{20}$

24. $\frac{13}{14}$

Challenge: $4\frac{1}{4}$ miles

Worksheet N, page 59

1. $\frac{7}{8}$ 2. $\frac{7}{12}$ 3. $\frac{6}{6} = 1$ 4. $\frac{8}{10} = \frac{4}{5}$

5. $\frac{9}{15} = \frac{3}{5}$ 6. $\frac{16}{10} = 1\frac{6}{10} = 1\frac{3}{5}$

7. $\frac{15}{8} = 1\frac{7}{8}$ 8. $\frac{12}{6} = 2$ 9. $\frac{29}{12} = 2\frac{5}{12}$

10. $\frac{28}{15} = 1\frac{13}{15}$ 11. $3\frac{9}{10}$ 12. $8\frac{11}{12}$

13. $2\frac{2}{3}$ 14. $6\frac{4}{10} = 6\frac{2}{5}$ 15. $6\frac{4}{7}$

16. $\frac{24}{30} = \frac{4}{5}$ 17. $\frac{5}{8}$ of a yard 18. $3\frac{3}{4}$ inches

Challenge: MARCH FOURTH (March forth!)

Worksheet O, page 63

1. $1\frac{1}{3}$ 2. $1\frac{1}{5}$ 3. $1\frac{3}{7}$ 4. $1\frac{3}{8}$ 5. $2\frac{2}{9}$

6. $2\frac{2}{6}$ 7. $3\frac{3}{4}$ 8. $2\frac{1}{9}$ 9. $\frac{1}{2}$ 10. $\frac{3}{5}$

11. $\frac{2}{5}$ 12. $\frac{2}{3}$ 13. $\frac{2}{5}$ 14. $\frac{1}{2}$ 15. $\frac{4}{9}$

16. $\frac{3}{8}$ 17. $\frac{5}{15}$ 18. $\frac{12}{16}$ 19. $\frac{45}{50}$ 20. $\frac{15}{35}$

21. $\frac{27}{72}$ 22. $\frac{18}{90}$ 23. $\frac{12}{54}$ 24. $\frac{40}{56}$

25. $\frac{4}{12} = \frac{1}{3}$ 26. $\frac{8}{9}$ 27. $\frac{13}{30}$ 28. $\frac{6}{12} = \frac{1}{2}$

29. $\frac{1}{9}$ 30. $\frac{3}{15} = \frac{1}{5}$ 31. $\frac{8}{12} = \frac{2}{3}$ 32. $\frac{29}{35}$

Challenge: $5\frac{2}{5}$ dollars or $5.40

Worksheet P, page 64

1. B 2. A 3. C 4. E 5. D

6. $\frac{1}{2}$ 7. $\frac{3}{6} = \frac{1}{2}$ 8. $\frac{2}{5}$ 9. $\frac{8}{9}$ 10. $\frac{1}{2}$

11. $\frac{3}{6} = \frac{1}{2}$ 12. $\frac{2}{5}$ 13. $\frac{1}{10}$ 14. $\frac{2}{1} = 2$

15. $2\frac{1}{6}$ 16. $2\frac{6}{9} = 2\frac{2}{3}$ 17. $2\frac{2}{4} = 2\frac{1}{2}$

18. $2\frac{2}{7}$ 19. $2\frac{4}{5}$ 20. $\frac{5}{1} = 5$

21. $1\frac{6}{8} = 1\frac{3}{4}$ 22. $\frac{8}{10} = \frac{4}{5}$ 23. $\frac{8}{8} = 1$

24. $\frac{10}{6} = 1\frac{4}{6} = 1\frac{2}{3}$ 25. $\frac{13}{12} = 1\frac{1}{12}$

26. $\frac{13}{8} = 1\frac{5}{8}$ 27. $\frac{33}{20} = 1\frac{13}{20}$ 28. $\frac{97}{56} = 1\frac{41}{56}$

29. $\frac{42}{30} = 1\frac{12}{30} = 1\frac{2}{5}$ 30. $\frac{77}{72} = 1\frac{5}{72}$

Challenge: $9\frac{3}{6} = 9\frac{1}{2}$ dollars or $9.50

Worksheet Q, page 65

1. $7\frac{4}{7}$ 2. $3\frac{8}{13}$ 3. $1\frac{16}{17}$ 4. $3\frac{7}{18}$

5. $2\frac{1}{18}$ 6. $2\frac{3}{13}$ 7. $1\frac{5}{18}$ 8. $2\frac{7}{15}$

9. $3\frac{1}{16}$ 10. $2\frac{3}{19}$ 11. $3\frac{9}{16}$

12. $2\frac{13}{23}$ 13. 3 14. $\frac{1}{3}$ 15. $\frac{1}{3}$ 16. $\frac{1}{23}$

17. $\frac{1}{3}$ 18. $\frac{1}{50}$ 19. $\frac{1}{3}$ 20. $\frac{9}{16}$ 21. $\frac{134}{7}$

22. $\frac{176}{5}$ 23. $\frac{481}{10}$ 24. $\frac{149}{4}$

25. $101\frac{1}{10}$ degrees 26. 12 days

Challenge: TWO-THIRTY (Tooth hurty)

Worksheet R, page 70

1. $\frac{5}{3}$ 2. $\frac{5}{2}$ 3. $\frac{11}{4}$ 4. $\frac{11}{5}$ 5. $\frac{10}{3}$ 6. $\frac{27}{10}$

7. $\frac{21}{8}$ 8. $\frac{15}{4}$ 9. $\frac{32}{9}$ 10. $\frac{27}{5}$ 11. $\frac{3}{5}$

12. $1\frac{3}{4}$ 13. $\frac{1}{5}$ 14. $\frac{3}{7}$ 15. $2\frac{2}{5}$ 16. $1\frac{5}{8}$

17. $1\frac{1}{2}$ 18. $1\frac{7}{10}$ 19. $1\frac{13}{24}$

20. $1\frac{9}{12} = 1\frac{3}{4}$ 21. $2\frac{1}{3}$ 22. $1\frac{3}{5}$

23. $1\frac{8}{12} = 1\frac{2}{3}$ 24. $2\frac{1}{3}$ 25. $3\frac{2}{3}$

Challenge: $5\frac{5}{12}$ hours

Worksheet S, page 71

The number formed is 100.

Worksheet T, page 72

1. 9 2. 6 3. 9 4. 8 5. 21

6. 56 7. 7 8. 35 9. 21 10. 35

11. $\frac{7}{8}$ 12. $\frac{5}{6}$ 13. $\frac{3}{4}$ 14. $\frac{3}{5}$ 15. $\frac{2}{3}$

16. $\frac{9}{10}$ 17. $\frac{5}{8}$ 18. $\frac{7}{8}$ 19. $\frac{9}{10}$

20. $\frac{9}{10}$ 21. $4\frac{1}{8}$ 22. $\frac{2}{4} = \frac{1}{2}$ 23. $3\frac{3}{5}$

24. $\frac{10}{6} = 1\frac{2}{3}$ 25. $\frac{8}{10} = \frac{4}{5}$

26. $2\frac{3}{5}$ inches   27. $14\frac{2}{3}$ feet

Challenge: 1. $1\frac{3}{8}$ pounds, 2. $2\frac{11}{15}$ hours,
3. $8\frac{1}{12}$ pounds

Worksheet U, page 76

1. $\frac{1}{12}$ 2. $\frac{1}{12}$ 3. $\frac{1}{5}$ 4. $\frac{1}{6}$ 5. $\frac{1}{4}$ 6. $\frac{1}{2}$

7. $4\frac{1}{3}$ 8. $4\frac{1}{2}$ 9. 4 10. $4\frac{2}{3}$ 11. $5\frac{1}{2}$

12. $1\frac{5}{24}$ 13. $1\frac{1}{2}$ 14. $1\frac{4}{21}$ 15. $1\frac{7}{30}$

16. $\frac{19}{20}$ 17. $\frac{1}{2}$ cup   18. $\frac{1}{4}$ tsp.   19. $\frac{1}{8}$ cup

Challenge: $1\frac{7}{8}$ quarts

Worksheet W, page 78

1. 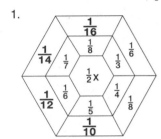   2.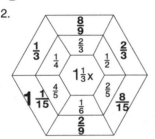

3. 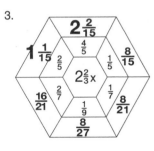   4.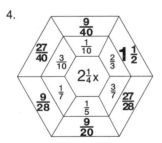

Challenge:

10	3	8
5	7	9
6	11	4

Worksheet X, page 80

1. $4\frac{1}{7}$ 2. $6\frac{2}{5}$ 3. $7\frac{1}{3}$ 4. $3\frac{1}{8}$ 5. $\frac{11}{2}$

6. $\frac{20}{3}$ 7. $\frac{17}{4}$ 8. $\frac{19}{2}$ 9. $1\frac{5}{12}$ 10. $1\frac{3}{5}$

11. $\frac{1}{2}$ 12. $1\frac{2}{3}$ 13. $1\frac{13}{24}$ 14. $\frac{1}{6}$

15. $\frac{3}{5}$ 16. $\frac{3}{8}$ 17. $\frac{5}{8}$ 18. $\frac{2}{5}$ 19. $\frac{7}{24}$

20. 2 21. 3 22. 9 23. 2 24. 2

25. $2\frac{2}{49}$

Worksheet Y, page 81

1. $\frac{1}{6}$ 2. $\frac{3}{10}$ 3. $\frac{5}{12}$ 4. $\frac{7}{12}$ 5. $1\frac{1}{2}$

6. $3\frac{1}{2}$ 7. $4\frac{1}{2}$ 8. $4\frac{1}{6}$

Perfect Numbers: $\frac{4}{5}$, $\frac{5}{7}$, $\frac{9}{10}$, $\frac{1}{2}$, $\frac{5}{12}$, $\frac{6}{7}$, $\frac{1}{3}$

Twenty-eight. Factors: 1, 2, 4, 7, 14. Yes.

Worksheet Z, page 82

1. $3\frac{1}{2}$ 2. $1\frac{4}{5}$ 3. $\frac{1}{4}$ 4. $3\frac{3}{4}$ 5. 1

6. 1 7. $4\frac{3}{4}$ 8. $2\frac{2}{5}$ 9. $2\frac{1}{4}$ 10. 7

11. $20\frac{1}{4}$ 12. $\frac{1}{2}$ 13. 7 14. 9

15. $1\frac{3}{10}$ 16. $\frac{7}{15}$ 17. $3 \times 2 \times 2$

18. $3 \times 9 = 3 \times 3 \times 3$ 19. $2 \times 27 = 2 \times 3$
$\times 9 = 2 \times 3 \times 3 \times 3$ 20. $2 \times 36 = 2 \times 2 \times$
$18 = 2 \times 2 \times 2 \times 9 = 2 \times 2 \times 2 \times 3 \times 3$

Challenge: width = $1\frac{1}{2}$ feet, length = $1\frac{2}{3}$ feet,
height = $1\frac{1}{4}$ feet

Worksheet AA, page 86

1. Circle these numbers: 2, 3, 5, 7, 11, 13

2. 3×2 3. $2 \times 2 \times 2$ 4. $3 \times 2 \times 2$

5. $3 \times 3 \times 2$ 6. 5×5 7. $3 \times 2 \times 3 \times 2$

8. $6 = 3 \times 2$ $8 = 2 \times 2 \times 2$
   L. C. D. $= 3 \times 2 \times 2 \times 2 = 24$

9. $12 = 3 \times 2 \times 2$ $9 = 3 \times 3$
   L. C. D. $= 3 \times 3 \times 2 \times 2 = 36$

10. $10 = 5 \times 2$ $6 = 3 \times 2$
    L. C. D. $= 5 \times 3 \times 2 = 30$

11. $10 = 2 \times 5$ $8 = 2 \times 2 \times 2$
    L. C. D. $= 2 \times 2 \times 2 \times 5 = 40$

ANSWER KEY

Challenge:

1. 3 = prime   6 = 3 x 2   10 = 5 x 2
    L. C. D. = 3 x 2 x 5 = 30
2. 8 = 2 x 2 x 2   10 = 2 x 5   5 = prime
    L. C. D. = 2 x 2 x 2 x 5 = 40
3. 15 = 3 x 5   10 = 2 x 5   6 = 3 x 2
    L. C. D. = 3 x 2 x 5 = 30
4. 18 = 3 x 3 x 2   10 = 2 x 5   3 = prime
    L. C. D. = 3 x 3 x 2 x 5 = 90

Worksheet BB, page 87

2. 9 = 3 x 3   6 = 3 x 2
    L. C. D. = 3 x 3 x 2 = 18
    Answer is $\frac{5}{18}$
3. 4 = 2 x 2   12 = 3 x 2 x 2
    L. C. D. = 2 x 3 x 2 = 12
    Answer is $\frac{4}{12} = \frac{1}{3}$
4. 4 = 2 x 2   3 = 3   6 = 3 x 2
    L. C. D. = 2 x 3 = 12
    Answer is $\frac{9}{12} = \frac{3}{4}$
5. 15 = 3 x 5   9 = 3 x 3
    L. C. D. = 3 x 3 x 5 = 45
    Answer is $\frac{8}{45}$
6. 25 = 5 x 5   10 = 5 x 2
    L. C. D. = 5 x 5 x 2 = 50
    Answer is $\frac{7}{50}$
7. 4 = 2 x 2   9 = 3 x 3
    L. C. D. = 2 x 2 x 3 x 3 = 36
    Answer is $\frac{13}{36}$
8. 14 = 2 x 7   10 = 2 x 5
    L. C. D. = 2 x 5 x 7 = 70
    Answer is $\frac{12}{70} = \frac{6}{35}$
9. 16 = 2 x 2 x 2 x 2   5 = 5
    L. C. D. = 2 x 2 x 2 x 2 x 5 = 80
    Answer is $\frac{21}{80}$

Challenge:

1. 8 = 2 x 2 x 2   3 = 3   9 = 3 x 3
    L. C. D. = 2 x 2 x 2 x 3 x 3 = 72
    Answer is $6\frac{41}{72}$

2. 15 = 3 x 5   25 = 5 x 5   9 = 3 x 3
    L. C. D. = 5 x 3 x 3 = 45
    Answer is $\frac{17}{45}$

Worksheet CC, page 88

1. $3\frac{1}{12}$ 2. $1\frac{7}{24}$ 3. $20\frac{11}{28}$ 4. $53\frac{13}{20}$
5. $5\frac{17}{21}$ 6. $58\frac{13}{25}$ 7. $5\frac{12}{35}$ 8. $58\frac{11}{24}$
9. $16\frac{23}{30}$ 10. $\frac{27}{35}$ 11. $1\frac{23}{42}$ 12. $1\frac{2}{5}$

Challenge: $6\frac{1}{4}$ cups raisins, 5 cups chocolate chips, $9\frac{3}{4}$ cups peanuts; 21 cups in all; $1\frac{1}{2}$ cups for each student

Fraction Assessment–Level 1, page 90

A. $\frac{2}{5}, \frac{2}{3}, \frac{1}{3}, \frac{5}{8}, \frac{1}{5}$ B. 5, 6, 9, 72, 56
C. $1\frac{1}{4}, 1\frac{2}{7}, 1\frac{3}{4}, 4\frac{1}{4}, 2\frac{2}{11}$
D. $\frac{5}{2}, \frac{15}{4}, \frac{17}{3}, \frac{29}{6}, \frac{38}{5}$ E. $\frac{2}{3}, \frac{9}{10}, 1\frac{1}{2}, \frac{1}{2}, 5\frac{7}{8}$
F. $\frac{1}{3}, \frac{1}{4}, \frac{1}{8}, \frac{1}{2}, \frac{2}{5}$ G. $\frac{1}{6}, \frac{1}{10}, \frac{8}{15}, \frac{5}{8}$
H. 2, 2, 4, $1\frac{4}{5}$

Fraction Assessment–Level 2, page 91

A. $\frac{7}{8}, \frac{1}{12}, \frac{7}{9}, \frac{1}{3}$ B. 30, 63, 35, 54
C. $5\frac{1}{7}, 3\frac{1}{5}, 3\frac{1}{3}, 4\frac{1}{6}$ D. $\frac{47}{3}, \frac{173}{8}, \frac{151}{4}, \frac{155}{8}$
E. $\frac{9}{10}, \frac{43}{50}, 1\frac{4}{7}, 1\frac{9}{50}, 7\frac{8}{15}$
F. $\frac{7}{30}, \frac{7}{30}, 1\frac{19}{24}, 3\frac{13}{18}, 8\frac{17}{20}$; G. $\frac{5}{12}, 1\frac{1}{8}, 2\frac{1}{8}$
H. 7, $2\frac{17}{20}, 2\frac{4}{9}$; I. 2 x 2 x 2 x 2 x 2 x 2,
5 x 5 x 3, 7 x 7 x 2